THE SOLAR ELECTRIC BOOK

THE SOLAR ELECTRIC BOOK

HOW TO SAVE $$$ THROUGH CLEAN SOLAR POWER

A PRACTICAL GUIDE

GARY STARR

INTEGRAL PUBLISHING
in association with
SOLAR ELECTRIC
1987

Printed in the United States of America

Published by Integral Publishing
in association with Solar Electric

Integral Publishing   Solar Electric
P.O. Box 1030   175 Cascade Court
Lower Lake, California 95457   Rohnert Park, California 94928

Computer-generated illustrations by E. R. Berland Graphic Design

Library of Congress Catalog Card Number: 87-080925
ISBN: 0-941255-45-X (hardcover)
ISBN: 0-941255-46-8 (paperback)
Printing (last digit): 10 9 8 7 6 5 4 3 2 1

CONTENTS

PREFACE

Dear Friend:

If I could spend just thirty minutes with you, you would become a user of solar electricity in one form or another. Why am I so sure? Because solar electricity makes sense. I regret that it is not possible to meet all my potential readers in person, so I have written this book to give you all the basics you need to know in order to put the sun's abundant energy to work for you—immediately and efficiently. Look at the facts and look at your priorities, and I am convinced that you will soon avail yourself of this great alternative source of free and clean energy.

There are all kinds of solar electric products available today that will improve your life or delight you and your children *and* that can save you hundreds of hard-earned dollars. Do you own a car? a boat? an RV? an airplane? a cabin or house? or a battery-operated electronic appliance, gadget, or toy? If your answer is Yes to even a single question, *solar electricity will save you money and time* and, in some cases, needless aggravation. (Just think of all those utility bills that are always higher than you would like.)

For many years the news about solar energy has been that it's not competitive with other sources of energy. This is no longer true. Also, I would like to point out that there is an important difference between conventional *solar energy* and *solar electricity.* Solar energy has proven its worth, though the public has been exploited by some unscrupulous entrepreneurs and is now rightly more cautious.

Solar electricity, also known as *photovoltaics,* which is what this book is all about, is quite different from the conventional hot-water panels that may be heating your neighbor's swimming pool. It is a relatively new technology requiring none of the complicated installations that are associated with solar heating. All the complexities of solar electricity are presolved for you by the manufacturers. You, the user, can enjoy an advanced power technology that is very simple to install and operate. What is more, solar electricity includes many applications that are definitely economical.

The facts and figures given in this book will effectively destroy the popular myth that all forms of solar energy are inefficient. Solar electricity is neither too exotic nor too expensive for widespread use. A growing number of people are recognizing this. You could be among them, benefiting from this benign form of alternative energy now rather than later.

You would be helping yourself and our planet, for solar electricity is a marvelous way of conserving the earth's natural resources and of taking responsibility for reducing environmental pollution. It is also a clean way of generating electricity because it doesn't add to the carbon dioxide or thermal pollution of our atmosphere.

If you are concerned about your budget, many solar electric products will help you keep your bank account balanced. If you are concerned about safety in your home, solar electricity is definitely your answer. If you are concerned about the future of Spaceship Earth, solar electricity is clearly the most sensible alternative. It can be considered an effective counter-measure to inflation, since many economists now agree that there is a direct relationship between high energy costs and inflation.

Moreover, economists have been warning us for some time now about the energy crisis, which is likely to worsen in the coming years. In March 1987, the highly reputed *Science* magazine featured an article that warned that "a future national energy crisis seems likely, probably sometime in the early to mid-1990s." Other experts are not as timid in their predictions. They tell us that the crisis is upon us already and that, if present trends continue, we will be facing a major catastrophe.

If we are wise, we will take such warnings into account when we make decisions that affect our and our children's future. Solar electricity can give you a large measure of independence in your energy requirements. Then, when the next oil crisis strikes and the "lights go out," you will not sit in the dark or face a heavy surcharge for "overconsumption" of energy, as was the case after the oil crisis of 1973-74.

This book gives you all the information and confidence needed to bring solar electricity into your life today. It's meant to be a practical guide that will answer all your initial questions. If you would like to know more about solar electricity and the products and services that are available, and if you wish to get in touch with the solar electric family around the globe, you may want to subscribe to my monthly publication *The Solar Electric Newsletter.* This will put you in touch with the latest news and views in the field as well as the friendly people who are currently using this source of energy. (See the advertisement at the back of the book.)

Also, if you're curious about how a solar electric cell works, I invite you to make use of the coupon at the back of the book. It will bring you a real working solar cell through the mail straight to your home—*free* and without any obligation.

Perhaps I can soon welcome you to the growing solar electric community. Or, if you are already a member of it, more power to your cells!

April 1987   Gary Starr

CHAPTER 1

SOLAR ELECTRICITY: HOW IT ALL BEGAN

The sun, we are told, has been illuminating our little corner of the universe for some 6 billion years. It is a massive ball consisting mostly of hydrogen, which acts like a thermonuclear reactor. The mass of this average star is estimated at 2 billion billion billion tons, a figure requiring 27 zeros.

The sheer quantity of hydrogen exerts a tremendous pressure on the sun's center, which translates into temperatures of around 30 million degrees Fahrenheit. This causes the atomic nuclei of the hydrogen atoms to fuse, forming heavier helium nuclei. In the process, a huge amount of energy is freed and hurled into space.

Every second, the sun emits 100 million billion billion (a figure with 26 zeros) calories of energy. The measure of the calorie is used by physicists as a standard unit for heat. It is equivalent to the amount of heat required to heat up 1 kilogram of water by 1 degree Celsius. This translates into some 380,000 billion billion kilowatts. By putting out so much energy, the sun consumes itself at the rate of around 4 million tons every second. That is to say, since its creation the sun has emitted the inconceivable amount of energy of some 20,000 billion billion billion billion calories. This is a figure with 40 zeros!

Only a thousandth part of the sun's total energy emission, consisting of different wavelengths, reaches our globe after traveling across 93 million miles of empty space. And only half that amount of energy penetrates the layers of cloud, haze, dust, and smog. This is not altogether a disadvantage, because only a portion of the total energy with which our planet is bathed is life sustaining. The atmosphere filters out most of the harmful radiation.

The sun's energy arrives 51 percent as heat (long, infrared rays) and 40 percent as light (the visible spectrum of energy). The rest is made up of the dangerous shortwave, ultraviolet rays.

Every square meter of the earth receives on average 1.395 kilowatts, making a total of 85,000 billion kilowatts. One kilowatt equals 1,000 watts, which is the fundamental unit of measurement of electrical power. But average figures are abstractions that may or may not be helpful. In practice, different regions of the globe receive different amounts of solar radiation. Here is a table for some U.S. cities:

Table 1: Average Annual Solar Radiation for Select American Cities in Million Kilowatt-Hours Per Acre

El Paso, Texas	9.5
Fresno, California	7.8
Miami, Florida	7.0
Salt Lake City, Utah	6.7
Lincoln, Nebraska	6.3
Cleveland, Ohio	6.1
Washington, D.C.	5.8
Boston, Massachusetts	5.2
New York City, New York	4.9

Farrington Daniels, in his book *Direct Use of the Sun's Energy,* gives this striking example. The desert of northern Chile, which is about 100 miles wide and 280 miles long, receives about 1,300 billion billion calories of heat every year. Daniels makes the point that this is roughly four times more than the total annual heat produced worldwide by the burning of coal, oil, gas, and wood.

Here is another way of looking at this: if as little as 1 percent of the solar energy reaching the Sahara Desert were converted into electrical power, the world would be supplied with electricity until the year 2000. Think of it!

Even more down to earth: as I have explained, the solar energy available in the sunniest part of a sunny day is around 1 kilowatt per square meter. This is enough to supply all the light bulbs in your house with electricity, or to run three television sets. At least, in principle, this is so. In practice, technology is able to convert only a portion of this energy into usable energy.

I spoke of our atmosphere as providing a buffer against the harmful spectrum of the sun's energy, but it does a great deal more. In conjunction with sunlight, the atmosphere has literally produced a seemingly infinite number of life forms. And this process is continuing even as you are reading this line. Without sunlight, you and I wouldn't even be here considering this matter. For life as we know it to develop at all, certain conditions have to be satisfied, and the presence of light is one of them. It is just as important as the air we breathe.

From earliest times, human beings have used the sun to make life on earth more comfortable. Our early ancestors dried hides in the sun, so that they could make suitable clothing out of them. They baked clay bricks in the sun to use in building their homes, keeping them cool in the summer and warm in the winter. Above all, since the neolithic revolution, our forebears used sunlight to grow food, just as we do today.

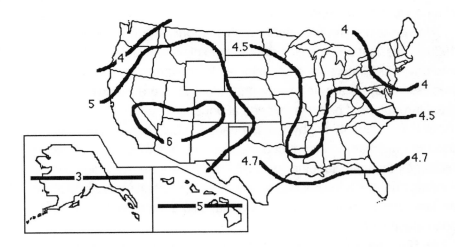

Figure 1: Average Hours of Sunlight Available in the United States

As we have seen, there is enough solar energy coming through the atmosphere—though it's only 5 percent of what we would encounter in space—to meet the energy requirements of all the nations in the world. In fact, it could supply every country with the same luxurious amount of electrical power that is currently consumed by the U.S. population, and there would still be 95 percent of it unused. The sunshine over the United States is, indeed, one hundred times in excess of present energy consumption levels.

Considering solar energy to be a possible source for meeting large-scale energy needs is a relatively recent development. The idea to use sunlight to heat up water for different purposes has been around for a long time, but it was only in the late eighteenth and early nineteenth centuries that some maverick scientists began to apply this knowledge to the newly invented steam engine. These men were ahead of their time, and their applications turned out to be too expensive for economical usage.

Then, a century ago, it was discovered that sunlight can produce electricity. This physical phenomenon, known as the photovoltaic effect, was first observed in 1839 by the French physicist Edmond Becquerel. He noted that a small voltage appeared when two metal plates were immersed in a conductive fluid and exposed to sunlight.

However, nothing was made of this astonishing discovery until the 1930s. Classical physics was unable to explain why solar cells worked, and power engineers were accustomed to converting heat into energy. It was essentially lethargy in science and industry that prevented Becquerel's breakthrough from being developed at once.

But then quantum physics happened. Max Planck, the father of quantum mechanics, introduced the idea that energy comes in discrete packages (quanta). Together with Albert Einstein, he revolutionized our thinking about how the cosmos works. In 1905, Einstein published not only his famous special theory of relativity, but also a remarkable study of the photoelectric effect.

In 1931, an article in *Popular Science* prophetically suggested that once solar cells were refined, these cells might power "a huge solar electric station at a cost no greater than would be required to build a hydro-electric station of the same capacity." This hasn't quite come true yet, mainly because the U.S. government chose to support the conventional power industry rather than solar electricity. In the meantime, however, many advances have been made in the solar electric field. These have made this form of energy commercially viable, and major breakthroughs in the mass production of even more efficient photovoltaic cells are expected in the very near future.

After the general acceptance of quantum physics, scientists began serious research on solar cells. In 1954, Bell Laboratories reported on the development of an improved solar cell with an efficiency of 6 percent, which for the first time made it practical to use solar electric technology as an energy supplier.

The real impetus for solar-cell technology came from the American space program. Perhaps you will remember the Vanguard I satellite, which was launched in 1958; it carried a solar array to power its radio. Solar cells in space proved more reliable than cumbersome fuel systems or batteries, and they were also the lightest per-watt source of energy. However, they had to be made by hand and so their costs were exorbitant at around $1,000 per peak watt, which is more than a thousand times the cost of producing electricity through fossil fuels, such as coal or oil.

The initial high cost of producing solar cells was primarily responsible for the limited interest in exploring their terrestrial applications. Nonetheless, some farsighted people recognized the tremendous potential of photovoltaics, or electricity produced from light. Thus the *New York Times* recommended in the late 1950s that "we ought to transfer some of our interest in atomic power to solar power." For various reasons, however, the American government has never allocated substantial funds for research into cheaper, more efficient solar cells. Instead it has plowed billions of dollars into the exploration of atomic power and the building of fission reactors. According to *The Hidden Cost of Energy,* a report published by the Center for Renewable Resources, the federal tax expenditures on nuclear power amounted to $9.9 billion in 1984. In addition, nuclear power received $2.3 billion in federal agency outlays and $3.3 billion in federal loans and guarantees.

As a consequence, there are now 106 nuclear reactors scattered throughout our country. Every year these reactors produce around 15 percent of the total domestic power. And what is worrying more and more people, they also produce 1,800 tons of radioactive waste every year. These materials are so toxic that they have to be safely stored away for 10,000 years. The only trouble is that scientists haven't yet discovered how to do this. Until recently, several countries around the world simply dumped their low-level radioactive waste into the ocean.

The fallout accidents of Three Mile Island and, more recently, of Chernobyl in Russia have given rise to much public concern about the safety of nuclear installations. Happily, fewer nuclear plants are under construction and several partially constructed plants have been abandoned—simply because they did not keep the original promise of plentiful power at low cost.

The dismantling, or decommissioning, of a nuclear plant that is no longer in operation is an extremely dangerous undertaking. It is also unbelievably expensive, running into tens or even hundreds of millions of dollars. Some of the really large reactors may cost as much as one billion dollars. To dismantle all the nuclear plants around the world, several hundred billion dollars will have to be raised. What a staggering figure!

Meantime the governments of the larger nations of the world continue to sponsor expensive research into producing nuclear energy through fusion (as in the H-bomb) rather than fission (as in the atom bomb and present-day nuclear power plants). Fusion is supposedly clean and safe, but this is a misconception. It too yields a certain amount of radioactive materials that will have to be transported and stored safely for many generations. At any rate, the technical problems involved in achieving controlled fusion seem at present insurmountable. To fuse hydrogen atoms, scientists have to recreate the conditions in the interior of the sun. Since no material can withstand the required temperature of millions of degrees, magnetic fields are used to trap this phenomenal energy. This may be exciting for curious physicists, but it also makes a big dent in the country's budget, claiming large sums that could be more usefully spent on, dare I suggest it, solar energy research and development.

Clearly, it is greatly desirable for the governments of the world to use their budgets more judiciously, given the demonstrated possibilities of such alternative energy sources as solar power. Instead of spending hundreds of billions of dollars, marks, and rubels on propping up the nuclear power industry and fueling the arms race, they could more profitably inject generous funding into solar energy research. As Sir George Porter remarked: "If sunbeams were weapons of war, we would have had solar energy centuries ago."

Despite the lack of government funding, the photovoltaic industry has grown steadily over the past thirty years. As mentioned earlier, in the late 1950s solar electric cells cost over $1,000 per watt to manufacture. By the late 1960s and early 1970s the production costs dropped to about $100 per watt (the cost of Skylab's solar electric wings). By the mid-1970s solar electric modules were down to $40 per watt, dropping another $20 at the end of that decade. While oil prices have dropped in the 1980s, the photovoltaic industry has also maintained its trend toward lower production costs.

Today, in larger quantities, solar cells cost around $6.00 per watt, and further reductions are expected in the near future. This has encouraged the industry to develop a variety of solar electric products.

From the small pocket calculator or battery charger to the megawatt power stations—photovoltaics is demonstrating its practicality. Recent breakthroughs in cost and efficiency are especially significant for the general consumer. Also, photovoltaics is becoming increasingly attractive as conventional energy prices rise. Contrary to common perception, solar electricity is no longer a futuristic technology. It has become a quiet revolution, which is reflected in the fact that over 200 million people own solar electric calculators.

The current market for photovoltaics is about $200 million per year and is expected to expand to up to $10 billion or even to $100 billion by the year 2000. What this means is that photovoltaics will be one of the fastest

Figure 2: Solar-Powered Stereo Equipment

Figure 3: Solar Electric Products Are Fun

growing industries over the next decade or so. Not surprisingly, Robert Metz in his syndicated column "Investing" has recently remarked that photovoltaics is the best market for any investor since solar "may soon begin to solve the world's energy problems for good."

The Stanford Research Institute predicted that by the year 2000, solar energy will meet from 7-13 percent of America's total energy needs. Moreover, the Union of Concerned Scientists (UCS), in a book-length report published in 1980, mentioned 20-25 percent and *recommended* 35 percent. The UCS figures are encouraging, as is the fact that there is a group of scientists who care about the runaway technology they have created. The issues behind such predictions are very complex, but the figures speak for themselves. Solar electricity, and solar energy in general, are here to stay. More than that, they will grow as an industry. Are you willing to grow with them into the twenty-first century?

The public demands products that are low in cost, maintenance free, and quiet. A reasonable demand. Solar electric products meet all three criteria. More than that, solar electric products are just plain fun. They intrigue us, and they kindle our creativity and our hope of access to free and limitless energy.

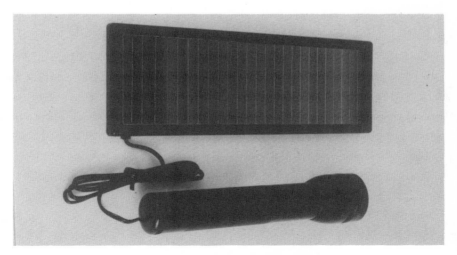

Figure 4: Solar-Powered Flashlight

CHAPTER 2

THE MAGIC OF ELECTRICITY

The modern world has an ongoing love affair with electricity. This is understandable. Electricity has brought us conveniences that were simply impossible before this force was harnessed: the light bulb, radio and television, portable batteries, automobiles, submarines, spacecraft, and, yes, electric train sets, and thousands of other pieces of technological ingenuity. These have become so much a part of our lives that we barely consider how recently they were invented.

Electricity has of course been around for a very long time, even before human beings became curious about it. It is, in fact, an integral part of the atomic world. The atom is composed of protons, electrons, and neutrons. Protons are positively charged, electrons negatively, and neutrons have no charge at all. In a typical atom, protons and neutrons form the core, with one or more electrons spinning around this nucleus.

Electrons are in the habit of jumping from one atom to another—but this fact is of advantage to us. For if enough of these wandering electrons can be made to jump at the same time and in the same direction, we get an electric current. This happens, for instance, when we stroke our cat or take off our sweater, or when we slide across the carpet and then touch the doorknob. There is crackling and, in the dark, we can see sparks of blue light. Miniature lightning bolts! What we see is the discharge of the electrons that have collected on the surface of the cat's fur, the sweater, and our own body. We call this static electricity because for a while the electrons just sit there, waiting to be released.

While air can conduct electrical currents, there are conductors—like copper, silver, or gold—that serve this purpose better. Electrons love to travel in these rather expensive metals, but they do so only if they can get a round trip. In other words, the electrical system in which they travel must form a closed loop.

This simple discovery was made as recently as 350 years ago, by the Englishman Stephen Gray. Even the ancient Greek philosopher-scientist Thales of Miletus puzzled over the curious property of amber, or fossilized tree sap. Amber, when rubbed, attracts light objects like grass, cork, or hair. Thales didn't really have an explanation, and so for centuries it was thought that amber (known as "elektron" in the Greek language) was magnetic.

William Gilbert, another ingenious English scientist, proved that this idea was wrong. He recognized, in 1600, that this was an electrical phenomenon. From then on, electricity became very popular with scientists. It took another 200 years before they discovered the difference between static and current electricity. The honor went to the Italian physicist Alessandro Volta. Now the stage was set for all the big electrical inventions—from batteries to electrical generators to the giant electrical power plants that generate several million kilowatts each. You may well ask, What about these power plants and the large utility companies that run them?

Need I say it, the conventional production of electricity has a number of drawbacks. The first is that it is expensive, costing three times more than small-scale power generation. The great inefficiency of the conventional power system is borne out by the fact that the world's electrical power industry is facing a debt of around $100 billion. The second drawback is the disturbing amount of environmental pollution that is caused by the big power plants, notably those operating on oil or nuclear fuel. If the big power corporations had to pay for rectifying these environmental damages, they would add many more billions of dollars to their existing debt. U.S. utility companies are already spending over $2 billion on pollution control.

Nuclear plants, which have given rise to unprecedented public protest, cost around $3 billion to build. This translates into $3,000 for every kilowatt of electricity they generate—and this does not account for their operating

Figure 5: Nuclear Power Is Hazardous and Expensive

costs and the hidden cost of radioactive pollution. The nuclear reactors in operation today in the United States supply around 16 percent of America's energy, yet they use only 2 percent of the energy locked in uranium, while wasting around 70 percent of the heat that is generated and creating huge piles of highly toxic materials.

The world's electrical power industry is clearly in bad shape. This is reflected in the price of electricity, which rose by 5 cents per kilowatt in the United States in the years 1973-85. Further rises are expected.

The inefficiency of the electrical power industry becomes particularly apparent when we examine the present-day "utility grid" system. The electricity that is delivered to most households in America and elsewhere is produced in a rather roundabout manner that turns out to be less efficient than most people believe or the utility companies would have us believe. In fact, it is estimated that the overall efficiency of the current electrical generation system is less than 30 percent.

The grid system is the outcome of earlier aspirations to centralize power generation for ease of modernization and national development (and for billing). This is how it works: heat created from the burning of fossil fuels (e.g., coal, oil, gas) or uranium (in nuclear reactors) is used to produce steam that drives massive turbines that then create an electrical current. This electricity, which begins at 4,000-25,000 volts, is next stepped up to 60,000-500,000 volts, so that it can be sent incredibly long distances to homes and businesses. But electricity doesn't arrive in that form at our power outlets.

Before we can use it, it has to be stepped down again by transformers, which reduce the voltage to 120-240 volts. Amazing, isn't it? All this complicated technology so we can flip a switch and turn on a light. Renowned energy scientist Dr. Amory Lovins has equated this process to "cutting butter with a chainsaw." Overkill is another way of putting it.

Figure 6: The Electrical Power Grid Is No Beauty

The ingenuity that has gone into the creation of power plants is admirable, but the conventional generation of electricity is associated with all kinds of undesirable economic and environmental consequences. These have all been talked and written about at length, though it seems a lot more talking and writing has to be done before the governments of the world will show more than a lukewarm interest in alternative ways of producing energy. Normally only governments have the resources needed to start building large-scale solar energy plants, boost research, and get the price of solar cells down. But since most politicians do not have the foresight to plan for the future, it is up to each one of us—the consumers— to create the demand for solar electricity so that private enterprise can take over.

As I have explained, the power companies have to do a lot of switching back and forth to get electricity to where it is needed. The standard electrical current for U.S. households is 120 volts AC or "alternating current." What this means is that the electrons flow first in one direction, then in the other at a rate of 120 directional changes (or 60 cycles) per second. This form of voltage is used because it can be generated on a large scale centrally and transmitted over long distances at high voltages with little power loss—unless of course the system happens to crash. Then the consumer of grid electricity sits in the dark or in the cold, until the fault has been corrected and the supply is restored.

In some cases, where a person lives in a remote area, he or she cannot even plug into the system without paying the utility company sometimes tens of thousands of dollars to get connected up. In my opinion, this is like burning bills. Not only has that person paid out dear money on a luxury, he or she has also given up his or her independence as a user of electricity.

Low-voltage DC or "direct current" systems can provide an economical and far safer alternative to grid-dependent electricity consumption. Solar electricity is an excellent means of generating direct current.

Solar electricity provides power on-site where needed, with no moving parts that can break or rust, and with an initial investment that requires no additional expense for fuel—ever. The fuel is absolutely free. No government is taxing sunshine yet! Nowadays, solar electricity is available for sums as low as $5.00 for individual photoelectric cells, $100 for RV or car systems, or $500 for home power generation. Once the purchase is made, the power generated is completely under the consumer's control, with no more blackouts, brownouts, or the otherwise inevitable rate hikes and international price wars.

For some applications, it is true, photovoltaics represents a higher initial investment, but if you were to amortize your capitalization over the life of the system, in many cases your additional monthly loan payments would be *less* than your regular utility payments. Also, unlike your utility bills, your house payments are tax deductible.

When were you last shocked about your high utility bill? And when were you last inconvenienced by a blackout? Remember July 13, 1977? That's when the whole of New York was paralyzed by a lightning-triggered blackout. On that hot summer night, 10 million residents sat in the dark, wondering and worrying. Some 2,000 people were arrested for looting. The larger buildings didn't even have water because no electricity reached the pumps. As for living on the twenty-eighth floor, you had to use muscle power. In hospitals, emergency equipment failed, and patients on respirators were kept alive by hand pumps. By 9 A.M. the following day, only a fraction of the power grid was restored. The mayor declared a state of emergency and urged people to stay at home. This was followed by New York's unplanned baby boom the next year.

Power outages are annoying but also instructive because we get to see just how dependent we are on electricity and the goodwill and efficiency of the power companies. In February of 1986, Sonoma and Marin Counties in California suffered a blackout on a particularly stormy night. People were sitting in their cold and damp houses, wondering when power would be restored to them. But at least one house, located in rural Sebastopol, had lights, heat, hot water, radio, and even television.

The owners had simply switched over to their solar backup system. This was reported in *The Pacific Sun,* a Marin County newspaper. But I didn't need to read this in any newspaper to know it was true, for the story was about my own home.

The journalists who wrote up the little story also didn't lie when they claimed that our solar electric installation gives us "comfort during inclement weather, security in an emergency, and savings on energy costs every day, rain or shine."

You could enjoy the same advantages. I was quoted in the article as saying: "The only reason people don't have solar now isn't cost, isn't technology—it's education. It's that simple. It's up to the consumer." This is true. It *is* up to the consumer, *it is up to you and me.*

Figure 7: A Solar-Powered Home

CHAPTER 3

THE SOLAR CELL— SUNLIGHT IN A GRAIN OF SAND

The basic building block of a solar electric system is the solar or photovoltaic (PV) cell. This is a semiconductor that converts light directly into electricity.

Most solar cells are composed of pure silicon into which impurities are introduced. Silicon is the second most abundant element on earth. Silicon is gained from quartz rock and various kinds of sand—yes, sand. The present manufacturing process of producing pure silicon and introducing the different types of impurities is complex and very expensive. Silicon usable as a semiconductor sells for about $30 per kilogram. However, despite the high initial cost for the high-technology equipment necessary to grow crystals, solar cells can now be produced economically. And as the demand for photovoltaic products increases, costs will continue to come down.

Although the mathematics of the processes occurring in a solar cell is rather complicated, the physical composition of such a cell is simple enough. When the solar cell is exposed to light, incoming units of light (called photons) strike the silicon atoms and get absorbed. The energy introduced into each atom by the invading photons causes electrons to be released. This process induces a random movement of electrons in the material. The difference in electrical potential between the negative (N) layer and the positive (P) layer then creates an electrical current.

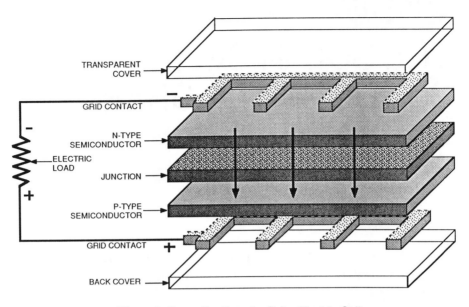

TRANSPARENT COVER

GRID CONTACT

N-TYPE SEMICONDUCTOR

ELECTRIC LOAD

JUNCTION

P-TYPE SEMICONDUCTOR

GRID CONTACT

BACK COVER

Figure 8: Cross Section of a Solar Electric Cell

This differential in electrical flow between the two sides is achieved through the addition of impurities to the silicon. Why silicon? The answer is simple: it has been found to respond best to the incoming photons. In other words, it promptly releases its electrons, creating the desired electrical current. One electron is released for every photon, providing the latter carries a certain minimum amount of energy. If it happens to carry a greater charge, still only a single electron is knocked out of each atom. The rest of the energy manifests as heat in the silicon material.

Most solar cells operate at an efficiency of around 16-18 percent and panels at around 8-12 percent. However, new developments are improving this efficiency. Only recently Stanford University produced a solar cell with a remarkable efficiency of 28 percent.

To turn silicon into a photovoltaic generator, silicon is grown as a crystal. It is melted down and then allowed to cool very slowly. In this way, the silicon atoms arrange themselves in a stable geometrical formation (a "tetrahedral" pattern). Scientists discovered that this arrangement is too perfect for maximum absorption of incoming photons, so they introduced minute quantities of other elements—boron and phosphorus. The addition of boron atoms depletes the silicon crystal of some of its electrons, whereas the addition of phosphorus atoms creates a surplus of electrons.

The boron-treated silicon layer, which has a positive charge, tends to soak up photons to make up for the deficiency. But the phosphorus-treated layer, which is negatively charged, always tends to get rid of electrons. The trick scientists discovered is to place a phosphorus layer on top of a boron layer. When one end of a wire is attached to the top layer and the other to the bottom layer, the electrons freed up through the onslaught of photons will travel from the phosphorus (N) layer to the boron (P) layer.

Between these two layers a curious phenomenon occurs. The "holes" of missing electrons in the boron layer attract surplus electrons from the phosphorus layer underneath, causing an imbalance in both layers. This zone of contact is known as the junction, which has a static charge. It is about one millionth of an inch thick. Only electrons with high energy can cross over from one layer to the other.

The amount of electricity produced depends on the size and efficiency of the solar cell. A cell is about one hundredth of an inch thick. This is no thicker than three human hairs piled on top of each other! Fragile as this artificial wafer is, it will function perfectly for many decades.

Some solar electric cells are made of polycrystalline silicon. These are composed of larger, granular crystals, which are easier and therefore cheaper to produce. These crystallites, or grains of crystal, can be grown several millimeters in length.

There are now also solar cells that are made of noncrystalline, or amorphous, silicon. This type of silicon does not have the internal symmetry of a crystal. Amorphous silicon cells are becoming inexpensive, though currently their efficiency of converting sunlight into electricity is lower, at around 9-11 percent, and 4-7 percent for complete panels. However, they are more sensitive to low amounts of light. This type of silicon cell is widely used in such small gadgets as watches and calculators.

Amorphous silicon cells are produced by the application of silica gas to a substrate such as glass or stainless steel. In this way, over one hundred times less silicon is needed. An ordinary silicon wafer is around 0.3 millimeters thick, whereas an amorphous silicon cell consists of a layer of 0.002 millimeters. What is particularly exciting is that the production cost of noncrystalline cells can possibly approach the cost of manufacturing the substrate—glass or stainless steel.

Semiconductors other than silicon are also being used and are likely to prove more efficient in the long run, though much research still needs to be done.

Single-crystal silicon claims around half the photovoltaic market. The production of this type of cell has been fairly stable at around 10,000 kilowatts for several years. But noncrystalline and also semicrystalline silicon cells are fast becoming serious competitors.

In the world of industry the race is on. There is a growing market for solar electric products, and many large corporations in the United States, Japan, and Europe are vying for a place under the photovoltaic sun. Until recently, the American industry was in the lead, but Japan is now claiming a larger share in the world photovoltaic market. European companies handle only about a fifth of the world production, but they have succeeded in opening up eager markets in third world countries.

The fastest growing market sectors in photovoltaics are stand-alone equipment (communication, village power, pumping, etc.) and consumer products, like battery chargers, calculators, and watches. In the period from 1980-86, stand-alone sales grew by 33 percent a year and consumer products sales by a whopping 75 percent a year.

The prospects for the photovoltaic industry are considered by economists to be good. This is worth remembering when you next consider investing in stock. Also, every dollar spent on solar electric products brings us closer to the goal of really low-cost photovoltaic products in every home.

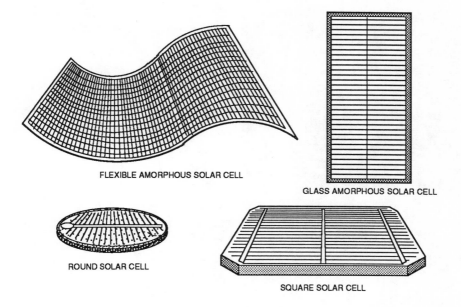

FLEXIBLE AMORPHOUS SOLAR CELL

GLASS AMORPHOUS SOLAR CELL

ROUND SOLAR CELL

SQUARE SOLAR CELL

Figure 9: Types of Solar Electric Cells

CHAPTER 4

SOLAR ELECTRIC PANELS—HOW IT ALL FITS TOGETHER

Figure 10: Solar Electric Cell

A single solar cell measuring 1" x 1" and putting out 1.6 volts at 70 milliamps has enough power to charge a single small battery or operate a musical greeting card chip. To operate larger equipment, a good many cells need to be joined to generate the necessary wattage. Such units are referred to as "panels" or "modules."

A small charge maintainer for a starter battery, for instance, yields around 1 watt. The same wattage is found in a solar-powered garden light with built-in rechargeable batteries. A panel large enough to power a gable fan should be big enough to generate 3-8 watts. To keep the deep-cycle batteries of a recreational vehicle in peak performance, a module generating around 3 watts is advised. A modest home power system, consisting of several modules, would have to produce around 250-300 watt-hours to handle light loads. A full-fledged home "power plant" would have to generate around 6,000 watt-hours a day.

In the early days of the photovoltaic industry, certain brands of solar panels had what were known as encapsulation problems. That is to say, the sealant or protective coating around each cell would turn yellow, warp, blister, or even crack. This is no longer the case with today's major brands, though some of the cheap modules manufactured in Asia suffer from these and other defects. Some of these imports consist of broken cells or "seconds" from U.S. manufacturers. These panels are pasted together by hand, using Asia's cheap labor force. While such patchwork modules can and do work, most of the ones I have inspected were not weatherproof and showed the kind of flaws that were characteristic of some of the early American modules. They are, therefore, best used for running toys or for doing experiments.

Generally speaking, however, there is little difference in quality between the solar panels marketed by the major manufacturers in this country and overseas. The main difference is in their specifications and efficiency ratings. Apart from this, manufacturers give different warranties, though it is important to remember that since solar cells and whole panels contain no moving parts, there is very little that can go wrong with them. I am the proud owner of one of the earliest prototypes of a solar-powered radio. It is now twenty-nine years old, but still works on its original solar cells.

The world's major manufacturers of solar cells and modules are oil companies. Oil and solar electricity? You might well ask. If you suspect a hidden conflict of interest, you would not be far off the mark. According to Ralph Nader's *Critical Mass Energy Project,* 70 percent of the U.S. uranium reserves are controlled by seven corporations, five of which are oil companies, including Atlantic Richfield (ARCO). My own policy is, to be quite frank, to favor the products of manufacturers that do not have oil as their principal commercial interest.

ARCO Solar panels are among the most widely used in this country. They also happen to be among the most efficient currently available. It is, however, my personal opinion that the advertising claims made for these collectors also apply to other brands. Recent tests by PG&E, a major utility company, showed that other brands, such as Sovonics, performed just as well, or even better.

Figure 11: Solar Electric Panels

Also, ARCO panels still use glass as the protective coating or as the substrate for the silicon cell. Unfortunately, glass is breakable and therefore liable to damage, and it invites vandalism.

ARCO Solar, Inc. is the largest photovoltaic manufacturer in the United States and is a wholly owned subsidiary of ARCO Solar Industries, which in turn is a subsidiary of the Atlantic Richfield Company, which happens to be one of the largest oil-producing companies in this country. In 1978, ARCO Solar bought out a small photovoltaic manufacturer that had been in operation for three years, and proceeded to install an automated assembly line. The company produces primarily single-crystal cells and modules, and their annual production totals a capacity of around 1 megawatt or 1 million kilowatts. Plans are underfoot for the construction of a plant manufacturing amorphous silicon cells, with an even more impressive annual capacity of around 20-50 megawatts.

ARCO Solar pursues intensive advertising and promotion, and is able to do so only because it is heavily subsidized by the Atlantic Richfield Company.

Solarex panels are another popular product. They are manufactured by Solarex Corporation, which became a subsidiary of the Standard Oil Company (Indiana) in 1983. Entering the photovoltaic industry in 1973, Solarex quickly assumed leadership in the field for technological innovation. Its most significant contribution has been the development of semicrystalline silicon cells.

Solarex panels are well made, which is what one would expect from a supplier to NASA, but they are also slightly higher priced.

Chronar panels, which are becoming popular, are somewhat less efficient. According to PG&E tests, these panels lose some power initially, but the loss tapers off after time. What this means is that you need to invest in a larger surface area to obtain the same wattage as from comparable panels by other manufacturers. These panels are commonly used in consumer products. They are produced by Chronar Corporation, which is a major supplier of amorphous silicon cells to the industry. This company seems to specialize in setting up joint ventures around the world.

Solec panels are of good construction and quality. They are manufactured by Solec International, Inc., a smaller privately owned company. It manufactures one of the largest modules available today, which is capable of generating 70 watts.

Sovonics panels use stainless steel rather than glass as a substrate, which makes this panel nearly unbreakable. For special applications, this brand can be ordered without a frame, which gives these panels the advantage of flexibility. Although Sovonics panels tend to be somewhat more expensive in small-scale systems, their high quality easily compensates for the extra cost.

This brand is manufactured by Sovonics Solar Systems, which is a wholly owned subsidiary of Energy Conversion Devices. This company, which is my personal favorite, was initially part owned by an oil company, but is now independent and continues its dedication to the commercialization of photovoltaic technology.

CHAPTER 5

PRACTICAL ELECTRICITY

Consumers of electricity owe a great debt to three gentlemen—the Scottish scientist Mr. James Watt (1736-1819), the Italian physicist Senior Alessandro Volta (1745-1827), and the French physicist-mathematician Monsieur André-Marie Ampère (1775-1836). Their genius has been immortalized in the three key technical terms connected with electricity—watt, volt, and amp. They stand for power, energy, and current respectively. This is how they are related to one another:

$$\text{watts (W)} = \text{volts} \times \text{amps}$$

$$\text{volts (V)} = \frac{\text{watts}}{\text{amps}}$$

$$\text{amps (A)} = \frac{\text{watts}}{\text{volts}}$$

An analogy may help you remember these relationships. Think of electricity as a water hose. Voltage is the water pressure. Amperage is the rate of flow: how much electricity moves through a circuit in a given time. And wattage is the power of the force that is the product of the voltage (pressure) times the amperage (flow rate).

You can calculate the voltage by dividing the wattage by the amperage. Similarly, you can find out the amperage by dividing the wattage by the voltage. So when you read, as I did recently, of a television worker who got jolted by 22,000 volts at 2 amps, you know that the wattage was 44,000. The poor man was thrown across the room, and was lucky to have escaped serious injury or death.

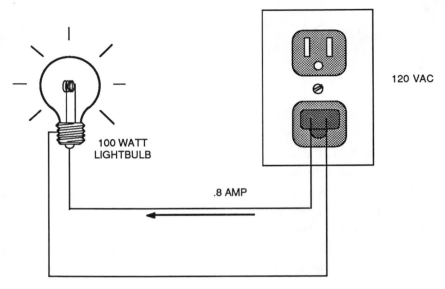

Figure 12: Typical AC Circuit

One other concept is necessary for understanding electricity, namely resistance. It is measured in ohms, which are named after the German physicist Herr Georg Simon Ohm (1787-1854). Resistance is a feature of materials that inhibits the flow of electricity. It is similar to friction. To use our analogy again, a large hose allows water to flow more freely than a small hose. Similarly, a small wire causes a lot of resistance to the flow of electricity. This is why the filament of a light bulb glows. Resistance, like friction, creates heat (emission of electrons).

The formula discovered by Ohm is:

$$\text{ohms (R)} = \frac{\text{volts}}{\text{amps}}$$

Let's use a practical example: you have a 100-watt light bulb, which is connected to a 120-volt outlet. Using the formula A = W (100) / V (120), we get 0.8 or 8/10 of 1 amp. Now we can calculate the resistance, using the formula R = V (120) / A (0.8), which gives us 150 ohms.

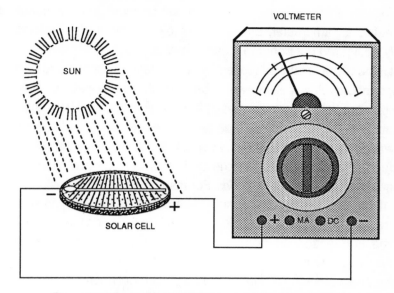

Figure 13: Measuring the Voltage of a Solar Electric Cell

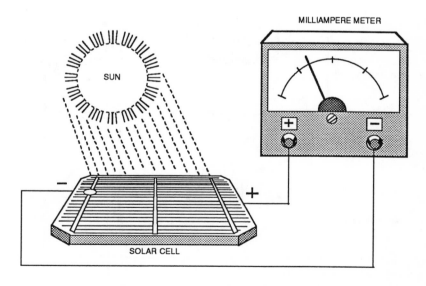

Figure 14: Measuring the Amperage of a Solar Electric Cell

Voltage, amperage, and resistance can be measured by means of a voltmeter, an ampmeter, and an ohmmeter respectively. Wattage can easily be figured out from the measurements obtained.

It is possible to take all these physical measurements in the case of a solar or photovoltaic cell. However, because the rate of electrical flow is so small, a special meter is required—the milliampmeter. There are 1,000 milliamps in 1 amp. Once voltage and amperage have been determined, the wattage can be calculated by using the standard formula (W = V x A).

Just as with batteries, voltage and amperage can be increased by connecting several cells *in series.* For instance, to produce 24 volts from two 12-volt solar cells, the panels must be wired together as shown in Figure 16. The positive (+) pole is connected to the negative (-) pole. If you get the connection wrong, something altogether different results. For optimal results, when connecting panels in series, the amperage should be similar and a blocking diode should be placed between the connections.

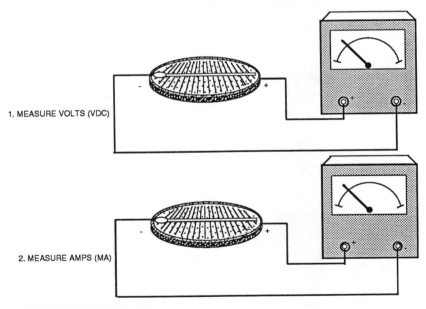

1. MEASURE VOLTS (VDC)

2. MEASURE AMPS (MA)

3. MULTIPLY VOLTS X AMPS = WATTS

Figure 15: Determining the Wattage of a Solar Electric Cell

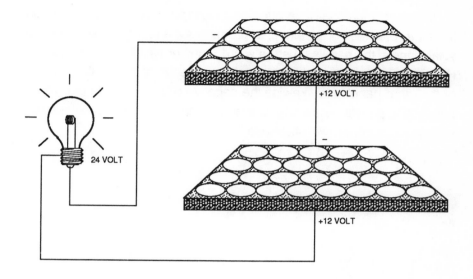

Figure 16: Connecting Solar Electric Panels in Series to Increase Voltage

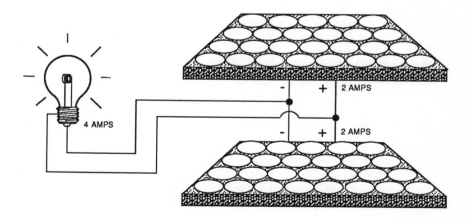

Figure 17: Connecting Solar Electric Panels in Parallel to Increase Amperage

In order to increase amperage (flow rate) in solar cells, the cells must be joined *in parallel.* For instance, to produce 4.2 amps from two 2.1-amp solar panels, the panels must be wired together as shown in Figure 17. Here the positive (+) pole of one panel is connected to the positive (+) pole of the other panel, and likewise the negative (-) poles are joined. When connecting panels in parallel, it is essential that the panels have the same voltage since panels will otherwise "even out" each other.

CHAPTER 6

PORTABLE SUNLIGHT: BATTERIES GALORE

Looking Back

Every year, millions of batteries are sold to an energy-hungry world. We use them in all kinds of electronic equipment and gadgetry—from toys to calculators, to watches, to flashlights, to radios and portable television sets, to cars. The average American spends over $100 every year on throw-away batteries. In the United States alone, the annual battery market is $4 billion. How many dollars do you spend for your portable electricity? When did you last buy batteries for your radio, flashlight, or tape recorder?

Do you know that you can now get a rechargeable battery that, coupled with a small solar-cell panel, could last for decades? And do you know that you could run even your portable stereo without any batteries at all, directly from a solar panel? A dry-cell battery, as found in flashlights or portable radios, is an amazingly expensive power source. It costs around $20-$100 per kilowatt-hour. The average grid electricity costs a 500th-3000th part of this. Coming in at around $6 per kilowatt-hour, solar electricity is a highly economical alternative to dry-cell battery power.

Batteries come in a great variety of sizes and voltages. The most common batteries are composed of dry cells, which carry a charge of 1.5 volts. They are made of a tightly sealed zinc container acting as the negative pole and a carbon rod in the center acting as the positive pole. Between them is a wet chemical paste that is separated from the zinc container by ammonia-soaked paper. So, dry cells are not so dry after all.

Figure 18: Assortment of Batteries

Still, a dry cell is drier than a wet cell, which contains a liquid solution of acid and water in which a copper rod and a zinc rod are placed. Between them the play of electricity occurs.

Another, more common, chemical generator of electricity is the lead-acid cell. In a so-called storage battery (a typical car battery), several such cells—three, six, or more—are connected. Each cell has a charge of 2 volts, and in serial connection, voltages of 6, 12, and more, can be achieved.

The chemical cells go back to Alessandro Volta's invention, the voltaic pile. It consisted of a number of alternating zinc and silver discs that were each separated by pasteboard soaked in saltwater. This chemical process of generating electricity is known as electrolysis.

Then there are fuel cells. Their fuel is also chemical, though it is neither dry nor wet, but gaseous. The two gases used are hydrogen and oxygen. These cells were developed for the space program.

Solar electric (photovoltaic) cells operate, as we have seen in Chapter 3, without a chemical medium. They generate electricity directly from the incoming light.

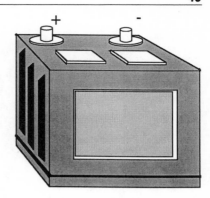

Figure 19: Car Battery

Car Batteries

The car or auto battery is designed to give a vehicle the powerful short burst of electricity necessary to get the engine started. This design feature is usually advertised as "cold cranking amps"—a great deal of amperage in a few short seconds. Once the engine is running, the battery is continuously recharged by the alternator. Under normal working conditions, a car battery is never discharged more than a few percent of its total capacity.

In fact, if a car battery is left sitting in a partially discharged or "deep-cycled" condition, it will soon die permanently. That is to say, you can't drain it and then hope to recharge it. A car battery that is allowed to sit idle for a period of time, without daily charge, often means a $40-$50 expenditure for a new battery.

While a big burst of initial electricity is useful in a motor-driven vehicle, this feature is not necessary or even desirable for a solar storage system for your home. You wouldn't want to put a race horse before a plow either. Rather, what is needed in powering a cabin or house is a battery that can supply a steady flow of electricity over longer periods. Using a car battery for solar energy storage would shorten its life greatly. It would be like leaving your headlights on a few times, which can easily destroy a battery.

Auto batteries are, therefore, not recommended for solar power systems. Only when extremely light loads are involved does it make sense to use this type of battery.

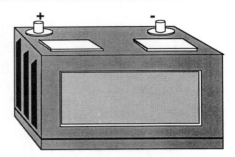

Figure 20: Deep-Cycle Battery

Deep-Cycle Batteries

Unlike car batteries, deep-cycle batteries are specifically designed to provide a steady flow of energy over longer periods. They can be completely drained before they are recharged. Hence they are an efficient way of storing and giving out power in a solar electric system. The most commonly available type of deep-cycle battery is that used in marine applications and RVs. They usually have a storage capacity of 85-105 amp-hours. Also, they are reasonably portable and will give you many recharges.

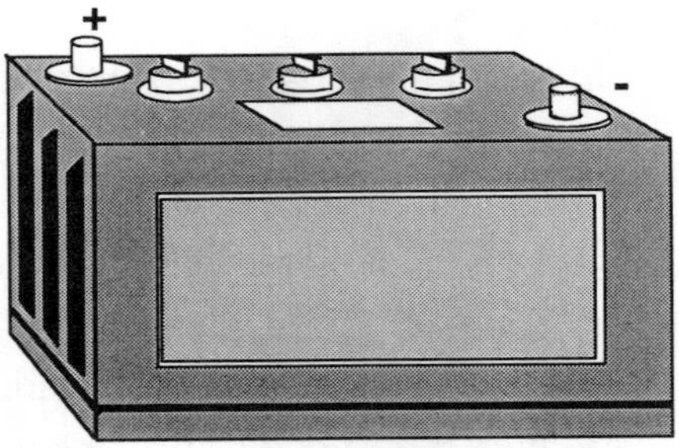

Figure 21: Golf Cart Battery

Golf Cart and Forklift Batteries

In situations where space is not at a premium, and greater storage capacity is desired, golf cart batteries are an excellent choice. These batteries have been used for years in many small to medium solar electric systems. Although heavier and more expensive than marine or RV-type batteries, they often have a storage capacity of 220-240 amp-hours and should prove a reliable power source for four to five years.

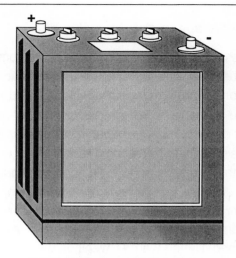

Figure 22: Commercial Duty Battery

Industrial Batteries

Industrial batteries are stationary deep-cycle batteries that are of extraordinary quality and often have a storage capacity of 300 amp-hours or more. These batteries are used, for instance, by telephone companies in various applications. Solar enthusiasts have reported acquiring used "telephone" batteries at bargain prices with excellent results. They are the heaviest of all deep-cycle batteries and also the most expensive. But the good news is that there are solar installations where these batteries have lasted for over eight years.

The biggest of these industrial-strength batteries carry a five-year warranty and are likely to last up to twenty-five years. Since they have a deep storage capacity, they often come in 2-volt cells and therefore must be wired together (in series) to yield a 12-volt current. They can weigh from 25 to 200 pounds each, but then they also have a storage capacity of up to 1500 amp-hours.

Perhaps the most expensive battery, which contains silver, is the type used in the NASA moon rover. It is also extremely reliable and stores the most energy for its size.

Scientists are working on "polymer" (plastic) batteries which have the same characteristics as silver batteries, but without either the cost or the weight.

Battery Storage

Batteries are the base of any solar electric home system. They serve as a buffer between the solar panels and the various appliances, providing power on overcast days or at night. Combined with solar electric modules, they are a reliable power source twenty-four hours a day.

The electrical storage capacity of batteries is rated in amp-hours. This rating is determined in each case by the manufacturer on the basis of the power that is discharged over a period of time at a certain temperature.

For example, if a battery is rated at 100 amp-hours, it will yield, say, 10 amps for 10 hours, or 5 amps for 20 hours, or 1 amp for 100 hours.

During cold weather, the electricity available from a battery is diminished. During warmer weather, by contrast, the available electrical power is increased, but so is the battery's natural self-discharge. Hence a battery needs a solar charge maintainer in the summer to offset this natural loss, and even more in the winter in order to offset both the self-discharge and the reduced flow of electricity. So, for best results, try to keep your batteries out of the freezing cold.

Table 2: Temperature–Electrical Current Relationship

Temperature in Degrees Fahrenheit	Available Electricity in Percent
80	100
32	65
0	40
-20	20

Battery Wiring

You can increase the net storage capacity of your system by combining batteries. In this case, they must be wired in parallel (positive to positive and negative to negative), as shown on the diagram below. This form of wiring is often used in RV applications or when two or more 12-volt batteries are used.

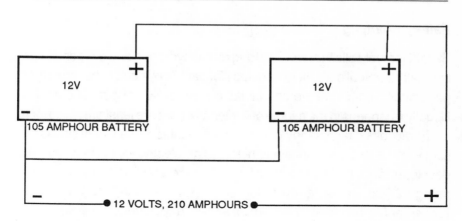

Figure 23: Parallel Wiring of Batteries to Increase Storage Capacity (Amp-Hours)

Net voltage can be increased by combining batteries in series (positive to negative), as shown in the following diagram. This kind of wiring is frequently needed when golf cart or industrial batteries are employed, since they are usually only 6 volts.

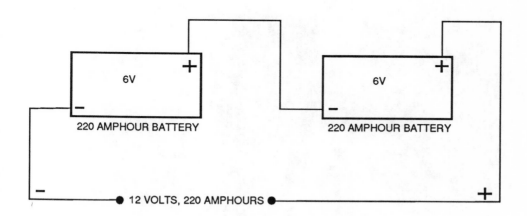

Figure 24: Serial Wiring of Batteries to Increase Voltage

Battery Charging

Except for fuel cells that work so long as the hydrogen and oxygen supply lasts, all chemical cells have a limited lifespan. They need to be recharged periodically. I am sure we can all tell a story or two about flat batteries. Usually it happens in the middle of winter when we are least prepared for it.

Photovoltaic technology can put an end to this. Battery charging is the single most practical application for solar electric cells. Industry has developed solar battery chargers for every battery size and capacity. Thus, chargers are available even for small 1.2-volt "AA," "C," and "D" *rechargeable* batteries. *The Solarizer™* is a particularly efficient and compact model.

Larger solar panels, which produce a minimum of 15 volts, are used for larger 12-volt DC battery storage systems. The storage batteries must of course be rechargeable, and are usually nickel-cadmium for the small sizes and lead-acid for the larger sizes.

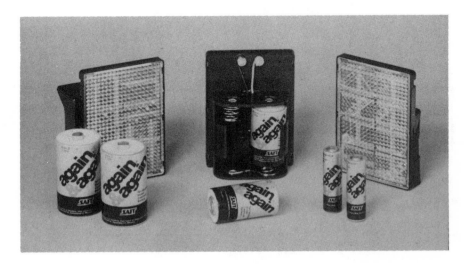

Figure 25: The Solarizer

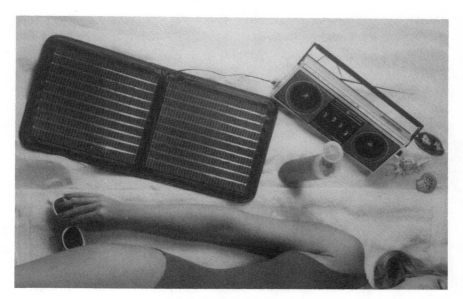

Figure 26: Solar-Powered Portable Radio

The Solarizer is especially designed to charge nickel-cadmium batteries (nicads). Although nicad batteries are higher in price than alkaline batteries, they should in most cases prove a better buy. Most nicads can be recharged up to 500 times and more.

Many first-time buyers are surprised to find their new nicads "dead." Nicads have the tendency to discharge themselves in storage, and so they are likely to need charging when you first purchase them.

A standard alkaline battery lasts longer than a nicad in noncommercial usage, but once you have emptied its charge, it becomes quite useless. A nicad may only last a third or a quarter as long as an alkaline battery per charge, but you can recharge and reuse the nicad over and over again.

For optimal economy, performance, and convenience, it pays to own at least two sets of nicads per solar charger, so that you can alternate between them. In this way, at least one battery is always in the charger on a windowsill or in any other sunny place, ready when you need to use it. When the battery in use becomes fully drained, simply replace it with the freshly charged battery and give it its rejuvenating turn in the solar charger.

Campers, owners of recreational vehicles, and backpackers find *The Solarizer* particularly handy. With it, they can always keep batteries charged while on the road (or in the wilderness). They never have to worry about damage due to battery leakage, and there is always a fully charged flashlight in case it is needed. Also, backpackers concerned with weight and availability of replacements for their batteries have in *The Solarizer* a convenient remedy for their battery blues. They swear by this device. So do motorists who will never be caught in the dark when a tire happens to go flat on them while driving along a poorly lit road at night.

Wherever you may find yourself, this portable battery charger can supply you with the energy needed to operate your radio, *Walkman™*, or camera. And you won't ever have to spend another penny on batteries!

James E. Culberson, a Florida adventurer-photographer, used three battery chargers during his 1987 expedition into the Likouala swamps in the People's Republic of the Congo, Africa. He went in search of a

Figure 27: "Close Encounter of a Solar Kind"

legendary creature known locally as Mokele-Mbembe. Eyewitness descriptions of this animal suggest that it belongs to the dinosaur family. Although Mr. Culberson didn't get a glimpse of the creature, native reports satisfied him that he was on the right track. And so he is preparing a new expedition into that treacherous 60,000-square-mile region.

On his next trip he plans to use a larger 12-volt battery to power depth recorders and spotlights for still photography and video footage. Since batteries quickly lose their charge in the extreme heat and humidity of the Congo swamps, he will again rely on solar chargers to supply the extra electricity needed. It is exciting to speculate that advanced photovoltaic technology might help bring to light an animal that is thought to have been extinct for millions of years.

Battery Maintenance for 6-, 12-, and 24-Volt Batteries

It is bad enough to run out of battery power for small gadgets (including children's toys arriving prettily wrapped but with dead batteries at Christmas and on birthdays). A dead car or motorcycle battery can be a lot more aggravating and serious. A dead battery in your boat, RV, or plane—just when you want to get away from it all—can be even more frustrating.

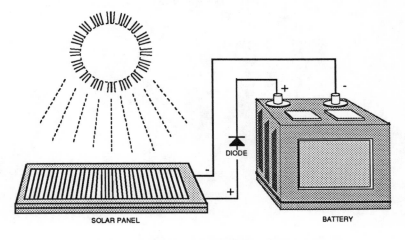

SOLAR PANEL   BATTERY

Figure 28: Solar Electric Panel Connected to a Battery

Or what about the heavy equipment you rented to get an important job done, only to find that the battery has died on you overnight?

Modern technology makes all these worries unnecessary. There are several kinds of solar-powered battery charge maintainers. With their help, you can now keep your 6-, 12-, or 24-volt battery fully charged, even if you leave it sitting for days, weeks, or even months.

The photograph on this page shows a private solar "fueling station." Its owner, Max Clouriter of Big Sandy, Minnesota, writes: "I have been using an array of panels delivering 1100-1200 watts to charge the batteries in a couple of converted EV Fiestas since September 2, 1984. I haven't had any trouble with these panels of cells. The only thing I do is adjust the angle of the array twice a year to catch more of the sun's rays. On most days it takes about 15 minutes of sunshine to replace the current in the batteries for one mile of travel."

Max's cars have a top speed of 66 mph, and accelerate from 0-35 mph in 600 feet. They have a range of up to 75 miles. The engine is a 20-hp

Figure 29: Private Solar Fueling Station

Prestolite DC motor, operating on sixteen 6-volt batteries.

Now, maintainers are not intended to recharge a completely dead battery. They are, rather, designed to prevent the battery from losing critical starting power. When a battery sits unused, it slowly discharges its power through a process known as sulfation. If a battery is allowed to sit uncharged for too long, it will be destroyed by crystallization caused by sulfation.

A solar-powered maintainer can, therefore, save you not only frustration and time, but also hard-earned money. By investing in a maintainer, which sells for about the same price as a new battery, you can prolong the life of your battery. Maintainers are like preventive medicine. Keep your battery optimally charged, and it will stay healthy and fit a great deal longer. And so will you, because you'll have eliminated one area of worry in your life.

Charge maintainers are also ideal for offsetting power drain caused by car computers, burglar alarms, radios, and cellular telephones.

Figure 30: The Maintainer

Maintainers typically come in two sizes. There is a 1-watt version for maintaining starter-type batteries and a 3-or-more-watt version for maintaining deep-cycle batteries. Deep-cycle batteries need about three times more energy for charge maintenance.

The Maintainer® comes complete with a 12-volt cigarette lighter plug. The dimensions are 4½" x 13" x ½", with a weight of 1 lb. The power output is 70 milliamps and 15.5 volts.

The Maintainer is placed on the dashboard and plugged directly into the cigarette lighter socket. If your vehicle doesn't have a lighter, simply connect the wires directly to the battery. This installation is easy enough, especially with the standard extension cords available today. You simply drill a hole where convenient so as to be able to connect the wire to the battery. However, since most cars are fitted with a lighter, the entire installation of your *Maintainer* can be done in a matter of seconds.

The Maintainer should be placed in a location, such as the dashboard, where it will benefit from direct sunlight. It can be left plugged in at all times, and there is no danger of overcharging the battery. *The Maintainer* contains a diode (one-way valve for electricity), so that it cannot drain the battery at night or on cloudy days.

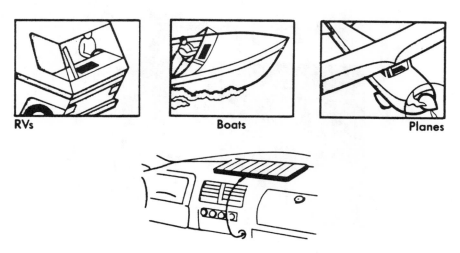

RVs   Boats   Planes

Figure 31: Positioning The Maintainer

Since these battery charge maintainers have no moving parts and are weatherproof, they should last for many years. And only one full hour of sunlight a day is needed to keep batteries fully charged.

Now, with this latest solar electric product, you can leave your vehicle standing unused for long periods of time, without having to worry about getting it started. The battery will be ready when you are to spark the engine at peak performance.

John Ahlf, of Ahlf's Enterprises in Santa Rosa, California, owns a backhoe service. He was plagued by dead batteries during the offseason. He now has a *Maintainer* for his backhoe and another for his truck. "It worked so well," he writes, "I bought two more. One is now on my tractor's battery, the other recharges a small portable light for my hay barn. It's amazing, it really works! During the day the solar panel charges the batteries, and at night I use the light to help me feed my horses. I feel that every piece of construction equipment should have a solar maintainer. Otherwise the operator is simply wasting time and money. At current rates, you could pay for *The Maintainer* what you would lose in a single hour of downtime. I'm so sold on solar electricity that I am now even selling solar electric battery chargers."

The Maintainer II is simply a more powerful model. It has the built-in capacity to recharge *drained* 12-volt batteries and to power larger radios

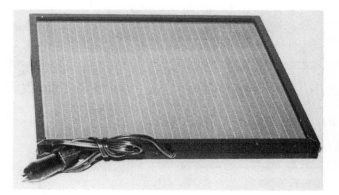

Figure 32: The Maintainer II

and even small portable black-and-white television sets. Its dimensions are 12" x 12" x ¾", with a weight of 3 lbs. The power output is 210 milliamps and 15.5 volts.

Twelve-volt battery maintainers are typically rated at 1-5 watts, with voltage outputs of 15-22 volts. The rated voltage output must be greater than the battery voltage in order to overcome the resistance of the battery and allow electricity to flow. Since a fully charged 12-volt battery can measure as high as 14 volts, solar charge maintainers must put out at least 15 volts to do the job.

With power output of this order, these solar maintainers can also recharge *small* rechargeable 12-volt batteries. Thus, they have been found useful in connection with computers, automatic bird and animal feeders, electric fences, alarm systems, and remote-site communications equipment.

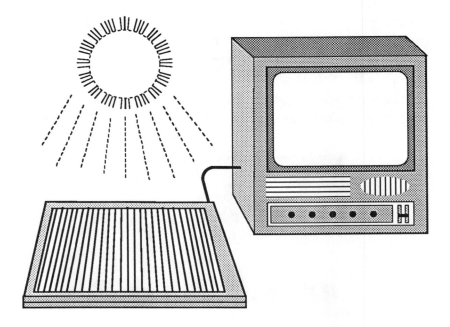

Figure 33: The Maintainer II Connected to a Television Set

Summer homes and outlying farm buildings in remote areas, where grid electricity is not available or unreliable, are prone to burglaries and vandalism. AHM Security Inc., a California company located in Orinda, has recognized the great advantage of an independent power source in this kind of situation. They now install solar-powered alarm systems for their clients and expect this service to become a new division within the company.

How to Test the Battery and Solar Panel

There are several devices to monitor the charge rate of your solar panel and the condition of your battery. The simplest device for checking the status of your battery is a charge indicator. This indicator connects directly to the battery and lights up depending on the charge status.

Another method for determining charge levels is by means of a hydrometer. This device indicates the specific gravity of the battery's electrolyte liquid. The gravity changes depending on the sulfate content of the electrolyte liquid. A scale on the hydrometer's float interprets this specific gravity in terms of the battery's charge status.

Figure 34: Hydrometer

You use it by opening a battery cap and siphoning the liquid into the hydrometer. The float in the hydrometer will read the liquid's specific gravity or charge rate. After the reading is made, simply return the liquid to the battery, but be careful not to spill any because it is acidic.

A voltmeter can also be used to determine the charge. However, not all voltmeters are sensitive enough to register the minute voltage changes that occur as a battery goes from "full" to "empty." A 12-volt battery has a full charge of around 14 volts that can decline to a low of 9 volts. Be sure to get a voltmeter with this kind of range.

A voltmeter can of course also be used to measure the voltage output of the solar panel. Perhaps the easiest meter to read is the digital voltmeter, which gives you a readout at the touch of a button.

In order to determine the current or amperage output of a solar panel, an ampmeter is necessary. It is important to get an ampmeter with the appropriate scale. A 35-watt solar panel, for example, will produce about 2.5 amps, while a 1-watt panel puts out about 70 milliamps. In the former case, the appropriate meter would have a full 5-amp scale, whereas the latter needs a scale of a full 100 milliamps. To obtain an accurate amp output reading, the panel must be hooked up to a motor or other efficient electrical appliance.

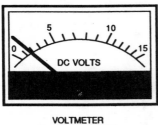

VOLTMETER
0 - 15 VOLT SCALE

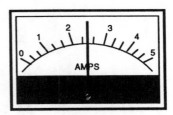

AMP METER
WITH 1 - 5 AMP SCALE

Figure 35: Voltmeter and Ampmeter

Figure 36: Solar Electric Panel Charging a Battery

Charge Controllers

When charging a 12-volt battery with a solar panel, one simply needs to connect the positive terminal of the solar panel to the positive terminal of the battery and the negative terminal of the solar panel to the negative terminal of the battery.

A blocking diode (electrical one-way valve) should be installed on the positive side to limit electrical flow to one direction and prevent any reverse flow from the batteries through the solar panel at night.

There are two types of diodes. The first is known as the P-N junction silicon diode for higher voltages; the other is the so-called Schottky barrier diode, which has a considerably lower voltage drop but is more expensive. Be sure that the diode has an amp rating higher than the panel rating.

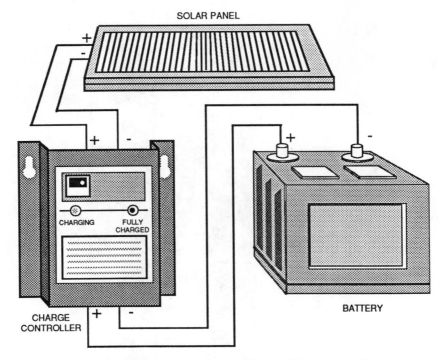

Figure 37: Solar Electric Panel with Charge Controller and Battery

If a panel rated greater than 3 watts or 210 milliamps is used to charge a starter-size or smaller battery, or if a panel rated greater than 8 watts (or 0.5 amps) is used to charge a deep-cycle battery, a "charge controller" should be connected between the solar panel and the battery. Since most charge controllers contain a built-in diode, it is not necessary to install a separate one.

The charge controller provides charging at maximum efficiency when the battery's charge is low. When the battery is fully charged, the charge controller provides a "float" charge similar to a solar charge maintainer to keep the battery at its healthiest charge level.

Some charge controllers also contain a fuse for protection, while others have built-in meters or indicator lights to register the charge rate of the panel and/or the status of the battery's charge.

CHAPTER 7

USING SUNLIGHT TO POWER MOTORS AND ELECTRONICS

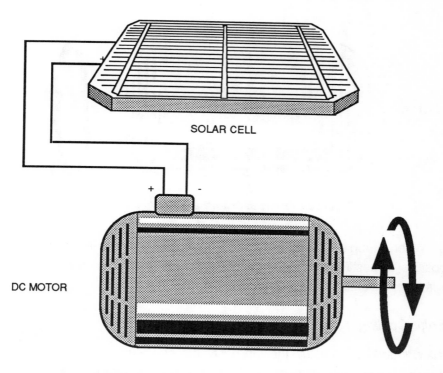

SOLAR CELL

DC MOTOR

Figure 38: Solar Electric Cell Connected to a Motor

Photovoltaics is definitely cost-effective for charging and maintaining batteries. But there are other money-saving applications of solar electricity that require no storage capacity at all. Here the solar panel is wired directly to a toy, calculator, radio, television, or water pump, and so on.

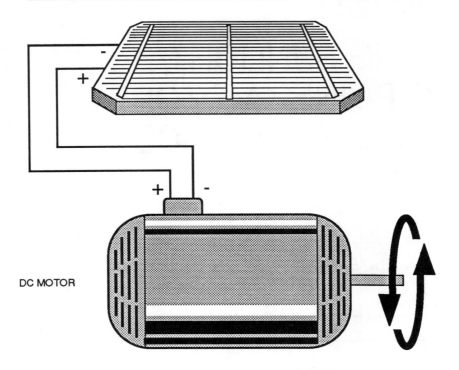

Figure 39: Reversed Polarity and Rotation

When wiring a solar panel to a DC motor, a negative-to-negative and positive-to-positive connection will cause the motor to run clockwise. If the wires are reversed, the motor will run counterclockwise.

Solar Cooling

It is wonderful to think that the sun, hot as it is, can help us keep ourselves and our homes or equipment cool. Solar-powered DC motors open up the possibility of solar electric cooling and the use of ventilating fans. The key is to have a really efficient DC motor that has adequate torque and is relatively inexpensive. Today a wide range of ventilation systems are available that operate directly from the sun. Everything from cooling a room to grain drying and greenhouse ventilation is possible through solar electricity.

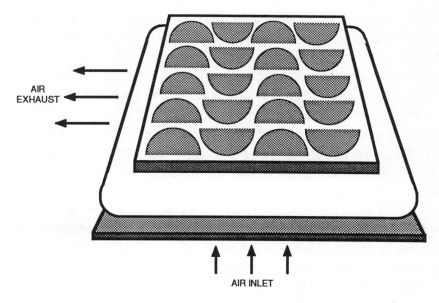

AIR
EXHAUST

AIR INLET

Figure 40: Solar "M" Ventilating Fan

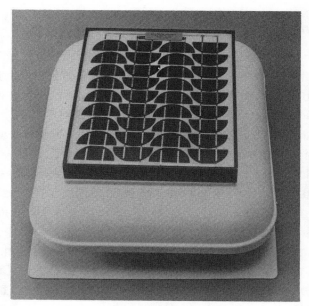

Figure 41: Solar "M" Ventilating Fan

Although solar-powered ventilating fans are usually higher priced than conventional fans, they have several advantages. To begin with, the solar electric modular "M" fan is easier and cheaper to install than its conventional attic-exhaust-fan counterpart because it requires no wiring. The solar panel is mounted on a sturdy housing and attached directly to the efficient DC motor and fan. Once installed, *Solar "M"™* has no further operating costs, since no commercial electricity is needed. The sun provides all the power necessary for cooling and ventilation—free. The hotter the sun, the more power the solar panel produces, and the faster the fan turns. This is all completely automatic.

This fan is self-contained and even comes with its own built-in flashing. A growing number of contractors and energy conservation specialists are installing solar-powered fans because of their ease of installation, moderate cost, and reliability. No wiring and no electrician are needed.

I will say more about solar cooling in the next chapter.

Marine Applications

There are many situations where, prior to the availability of solar-powered fans, it was impractical or too expensive to install any kind of ventilation at all. Solar-powered fans are ideal when cooling or ventilation is desirable but no conventional power supply exists. Thus, they have been used effectively in houseboats and other marine applications, as well as animal barns and trailers, greenhouses, and outhouses.

Most small motor boats and sailboats are only occasionally used on weekends during the summer recreation season. The rest of the year they generally sit in storage or are docked. What this means is that many boat owners have to periodically haul the boat battery home or to a charging station. This translates into time, money, and needless inconvenience.

A solar electric charger, like *The Maintainer* (see Chapter 6), is the cheaper and easier alternative to this haulage method. Its trickle charge prevents the battery from sulfating through nonuse and maintains it in

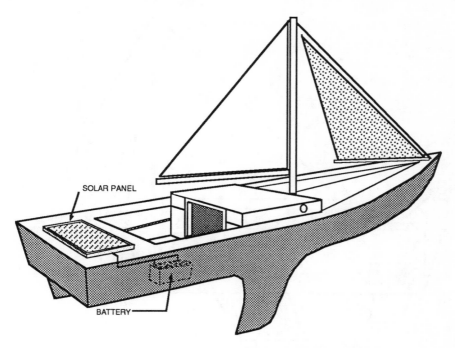

Figure 42: Boat with Solar Electric Panel and Battery

tip-top shape throughout the year. This device also dramatically extends the life of the battery.

Larger vessels with heavy energy demands often have multiple deep-cycle batteries. Sometimes boat owners install a backup diesel generator. But generators are costly pieces of machinery to maintain, and they bring their own problems with them—noise, stench, space, spare parts, and fuel consumption.

Where the energy requirements are higher, larger solar panels can provide the power necessary for all the boater's electrical needs. Photovoltaics can produce the energy for an extended cruise or even for full-time occupancy. The size of the panel array is obviously to be determined by the electrical usage. The current needed by some standard marine devices is as follows:

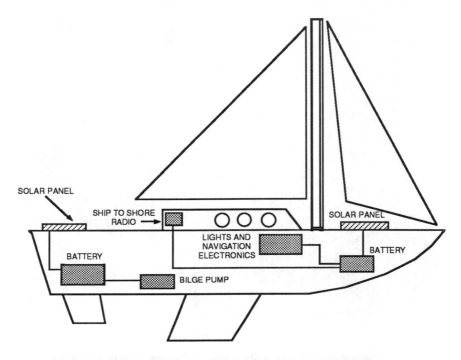

Figure 43: Multiple Usage of Solar Electric Panels for Marine Vessels

Bilge pump	0.3-2 amps
Depth sounder	0.4 amps
UHF receiver	0.5 amps
UHF transmitter	1.1 amps
Running lights	0.5-1.5 amps

The current draw of other 12-volt DC devices can be determined from the product specifications or instruction manuals, or by measuring the device with an ampmeter. If the wattage is known, the amperage can be calculated by dividing the wattage by 12 volts. The current draw of a number of other appliances is given in Chapter 10.

To determine solar panel requirements, multiply the current in amps by the hours of expected use per day for each device. The total consumption of amp-hours per day then must be divided by the number of hours of available sunlight. This gives you the number of amps your solar electric system will have to produce. Here is an example:

Step 1: Add up the number of amp-hours for all devices (20.5)
Step 2: Divide by the number of available sunlight hours (5)
Step 3: Determine the total number of amps needed (20.5/5 = 4.1)

What this means is that you would need solar panels with a 4.1-amp capacity, which, expressed in wattage, is 63.6 watts (4.1 amps x 15.5 volts). This power need can be handled by a single large module or several smaller ones.

Once installed, the photovoltaic panel performs for you, quietly and efficiently, day after day. All you have to do is enjoy the breeze and sunlit blue sky while crisscrossing your favorite body of water.

Figure 44: Solar Peace

This is exactly what Robert De Haan is doing in his spare hours. He writes: "I've been sailing for forty years and have owned sailboats for about twenty years. A year ago, my son-in-law acquired a *Maintainer II* to charge the deep-cycle battery aboard his engineless twenty-footer. It worked so well and effortlessly that I purchased two larger marine panels from Solar Electric for my Whitby 42. The 10-watt panel, mounted permanently on a hatch garage, keeps the engine starting bank fully charged. The 30-watt panel provides power to the other bank, which I use for lights, radio, depth-sounder, bilge pump, etc. This panel stores below a bunk and can be positioned almost anywhere topside for best sun angle.

"Before solar power, I had to run the engine several times a day to charge the batteries. Now I expect our days in San Joaquin Delta will be much quieter—with much less engine time. And the engine starting panel works whenever there is sun. So, no more worries about having weak batteries. With the flexible marine panels, solar power has finally arrived for the yachtsman. I recommend them for every boater who has batteries aboard."

CHAPTER 8

SOLAR ELECTRICITY FOR RECREATIONAL VEHICLES

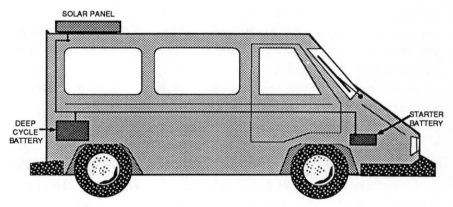

Figure 45: Solar-Powered RV

There are about 9 million recreational vehicles (RVs) in this country alone. RVers demand carefree, quiet, independent living. Solar electricity furnishes all three, and more. RVers typically want to "get away from it all," yet many begin their holidays by replacing an expensive dead battery. On the road, it is the sound of the gas-guzzling generator that has them sitting on edge. RVers tend to be concerned with energy consumption and energy conservation.

A solar electric array turns the RV into a self-sufficient energy system that includes the efficient, dependable, and quiet charging of batteries.

RVers combine their quest for peace in remote regions with a healthy respect for comfort. But the kind of comfort that we have grown accustomed to is impossible without electricity. Since "getaway places" have no commercial electrical hookups, most RVers have in the past invested in noisy generators to meet their energy needs. Solar battery chargers make this method obsolete. RVers can now travel to the remotest backcountry and still enjoy the peace and quiet of the wilderness. They can park indefinitely, using the silent rays of the sun to keep the batteries fully charged.

RVers and motor-home and travel-trailer owners share many of the boat owners' problems. They do not, however, have the option of hiring dock services to keep their batteries charged in the off-season. And without weekly trips to the storage garage to set up the conventional AC charger, a good many RVers start the holiday season with a new battery added to their shopping list.

As mentioned in Chapter 5, it is easy to install a charge maintainer to keep a battery in peak performance by trickle charging it with daily doses of solar energy. If the RV has more than one battery, a larger solar panel may be needed. I recommend that you provide 1 watt per starter battery and 3 watts per deep-cycle battery. So, if your RV has two deep-cycle batteries and one starter battery, you should have solar panels generating around 7 watts. It is usually more economical to purchase one larger panel than several smaller ones. A 7- or 8-watt panel would do the job perfectly in this case.

A battery charger of this capacity is included in *Trailercool™*, which is a large ventilating and charging system designed specifically for RVs. *Trailercool* fits into any standard 14″ x 14″ roof vent and generates enough solar electricity to either aerate and cool the RV or charge its 12-volt batteries. Although many RVs contain 12-volt ventilator fans, most owners don't run them because their inefficient motors are a drain on the battery. *Trailercool* contains a highly efficient motor that can run on free solar power.

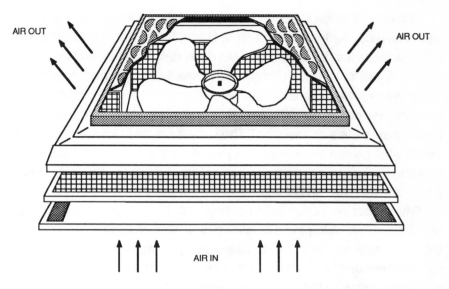

AIR OUT

AIR OUT

AIR IN

Figure 46: Trailercool

Figure 47: Trailercool

"Trailercool made a world of difference when we took our trip to the desert in 1986," writes Bruce Bryer of Dixon, California. "The brighter the sun was the faster the fan turned!" Mr. Bryer delights in the fact that when his fifth-wheel trailer is just sitting in the driveway, he can watch the charge meter move into the green zone every day as the sun recharges his batteries. He continues: "On our last trip we met some RVers who had larger solar panels on their roof. They never needed to run a generator. That's what getting away from it all should be like—peace and quiet and freedom from messing with noisy generators and gas . . . It simply doesn't make sense to own an RV and not use solar power."

With proper ventilation, as is possible with *Trailercool,* the use of an air conditioner can be kept to a minimum or is altogether unnecessary. The trick is to prevent the RV from overheating in the first place. Since air conditioners devour electricity, reducing or eliminating their use is a big step toward making the RV energy independent.

Another highly efficient and practical ventilation device is *Sun-Mate™*. This is a small solar electric ventilator that fits into the RV's refrigeration duct. Most RV refrigerator units work by expelling heat through the refrigeration duct. A small fan that assists this process will allow your refrigerator to operate more efficiently. Since the sunnier and hotter it is the more cooling is needed, a solar electric fan is the perfect answer. For the more sunlight there is the more power the solar panel will produce and the faster the fan will turn. *Sun-Mate* can save you dollars on electricity or propane. It will also help keep your food cooler.

To determine what size panel you need to make your RV energy independent, you simply calculate the approximate power consumed on an average day. In this way your solar electric panel will help you eliminate the use of an expensive and noisy generator. Proceed in the step-by-step manner described for marine applications in Chapter 7.

In many instances, one 2-amp panel and one storage battery per person will be adequate. If more power is needed, you can always add further panels to produce more electricity. It is recommended that any panel rated at more than 0.5 amps be installed with a charge controller to

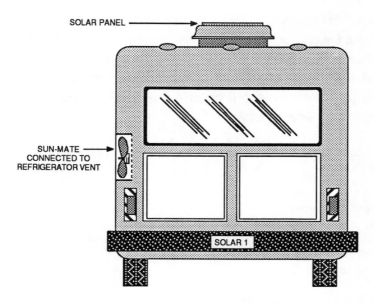

Figure 48: Sun-Mate Connected to Refrigeration Vent

Figure 49: Sun-Mate

ensure that you won't overcharge your battery. This installation can be done very easily. Simply connect the positive and negative wires to the charge controller and then to the battery.

If you had eight hours of sunlight, a 2-amp panel would yield 16 amp-hours (AH) per day. This panel is capable of powering the conservative load given in the following table:

Table 3: Load Capacity of a 2-Amp Panel on Eight Hours of Sunshine

	Appliance	Amps	x Hrs/Day	= AH/Day
1	15-watt fluorescent light (over kitchen)	1.1	5	5.5
2	8-watt fluorescent lights (over bed)	0.7	3	2.1
1	9-inch black-and-white TV	0.9	4	3.6
1	water pump (total running time)	0.6	2	1.2
1	cassette deck	0.8	2	1.6
	Total			14.0

As you can see, if your electricity consumption is conservative, one 2-amp panel can provide sufficient energy to power your RV's basic appliances, with a little extra to spare. The surplus power can be stored in the battery for the occasional use of a vacuum cleaner, a blender, or extra lighting.

If your energy requirements are greater, you will need additional panels. But you can always add these later on. The important thing is to get started. Once you see what solar electricity can do for you, you will wonder why you didn't solarize your RV earlier. If you own an RV, it simply doesn't make sense not to own a solar electric system. You won't ever have to deal with a noisy, messy generator again—and you can begin to save on those exorbitant hookup fees in trailer parks.

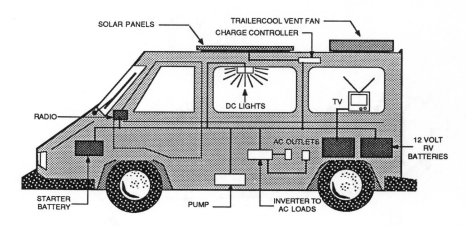

Figure 50: Typical Solar Electric Battery Charging System for RV

If you buy your solar system on credit (using your Visa or MasterCard, etc.), your solar electric installation can pay for itself instantly through a positive cash flow. You will save money on gas and batteries right away. Moreover, the savings you will realize by avoiding an average of two daily hookup fees per month will be greater than your monthly credit payments. Once your loan is paid off, the sun will be producing free dollars every day for your travel and leisure enjoyment.

What if you happen to enjoy your color television, microwave oven, and air conditioner? Can solar electricity still work for you? YES. Even if you belong to the big-time consumers of electricity, you can still harness the sun's energy to turn your RV into an independent power station. You will need more panels, but it can be done and, indeed, has been done.

First, let's look at your air conditioner. Most frequently, you will need an air conditioner when your RV sits in the sun all day heating up without proper ventilation. With an *efficient* DC fan, such as *Trailercool,* the inside of your RV will remain at a comfortable temperature with only occasional use, or possibly even without the use, of an air conditioner. If you do, however, want to run the air conditioner for long times, you may occasionally have to run your generator or use an AC electrical hookup.

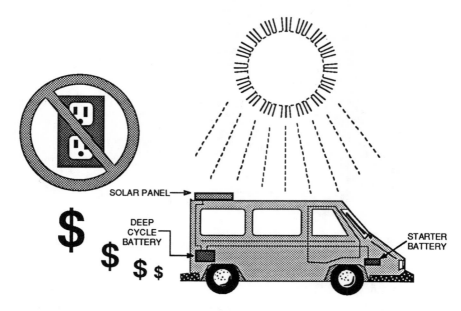

Figure 51: Hookup-Free RVing

But if you use 120-volt AC appliances, like the air conditioner, more judiciously, you may only need an *inverter*. An inverter changes your 12-volt DC battery power to 120-volt AC current. This quiet electrical device enables you to operate 120-volt appliances without the use of a generator or an AC hookup.

It is, however, important to note that most 12-volt DC appliances are generally much more energy efficient than 120-volt AC appliances. Today there are 12-volt blenders, tools, fans, hair dryers, razors, and dozens of other items. If you shop carefully, you may eliminate the need for an inverter.

Should you need an inverter, however, make sure you size it correctly, after checking the *maximum wattage* it must provide. Inverters are sized in watts. All you have to do to determine the right kind of inverter for your needs is add up the total number of watts that will be used at any one time.

Here is the approximate wattage of typical AC appliances in an RV:

Electric razor	30
Curling iron	40
12-inch black-and-white television	100
Fruit juicer	100
Computer	150
Electric blanket	170
Electric typewriter	250
Coffee maker	600
Corn popper	600
Iron	1100
Toaster	1100
Microwave oven	1200
Hair Dryer	1200

Installation

Determining the size of the panels you need to operate all the different appliances is not really difficult. Neither is mounting the panels on the roof of your RV. Find a spot where they can be firmly secured and where air conditioners and travel racks won't block out the direct rays of the sun.

The solar panels should be mounted slightly away from the roof, so that air can circulate beneath them, cooling their underside, which increases their performance.

It is best to use L-brackets, which should ideally be attached, wherever possible, to roof rafters. To prevent leaks, apply roof cement or silicon caulking around and under the brackets.

CHAPTER 9

BACK IN THE QUIET COUNTRY

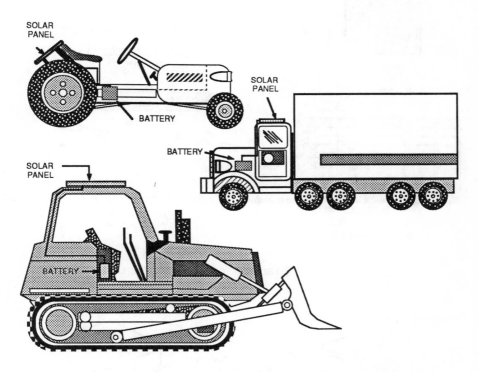

SOLAR PANEL

BATTERY

SOLAR PANEL

BATTERY

SOLAR PANEL

BATTERY

Figure 52: Farm Machinery for Solar Electric Conversion

Solar electricity is almost like the goose that lays golden eggs when it comes to farms and ranches. Since they are located in open country where sunlight and space are no problem, they are ideally suited for utilizing this form of energy production. Also, their rural location often means that regular grid electricity is not available or uneconomical, especially for outbuildings.

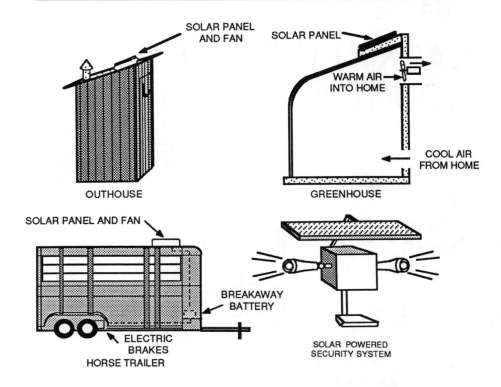

Figure 53: Solar Electric Farm Applications

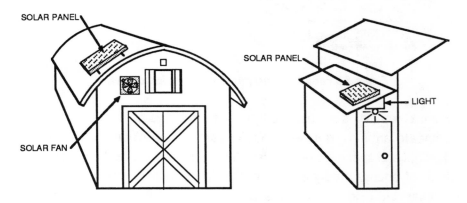

Figure 54: More Solar Electric Farm Applications

There are a multitude of single-use applications on farms or ranches where solar electricity is clearly advantageous. For instance, almost every farm or ranch has a tractor, an extra vehicle, or heavy machinery that does not have daily use. All of them require starter batteries, and all of them should have a solar charge maintainer. The savings in time and battery replacements would pay for the cost of a solar maintainer in a very short time.

Solar electricity is also the optimal answer for many outbuildings, animal barns, or storage sheds that require small amounts of electricity. Solar electricity can easily and economically provide power for light, ventilation, or a security system.

Solar Fence Charging

Decades ago, it was discovered that a small, harmless jolt of electricity could be used to deter animals from straying. Ranchers and farmers have put this discovery to good use. By wrapping an inexpensive charged wire around the corral of the penned livestock, they are able to restrain the animals without high initial investment and with little maintenance.

Figure 55: Solar Electric Fence Charger

Unfortunately, however, the area to be fenced in is often nowhere near any electrical outlet. So, in the past, farmers or ranchers have improvised and used batteries to power their fence chargers. But this has not proven very practical, since batteries need routine recharging or replacement, which is yet another chore to be added to an already full day.

A solar electric panel can do away with this expense and inconvenience. A small panel can provide enough power to keep maximum shock capacity on a fence at all times, with no operating interruptions. It even does away with operating costs. Most solar fence chargers are designed to charge the battery even on overcast days and are sized to be operated for up to three whole weeks without sunlight. They cost around $150, and solar retrofit kits for existing chargers, containing a rechargeable battery and a small solar panel (such as *The Maintainer),* are available for about $75.

Water Pumps

A solar-powered pumping system for irrigation or drinking-water requirements is an economical alternative to other methods where no utility power is available. Advances in motor technology and reductions in the cost of photovoltaics have opened the market for this particular application. Hence solar-powered pumping systems are being installed in many parts of the world.

In this application, solar electricity is definitely competitive with windmills and gas generators. Wind is notoriously unreliable, whereas gas generators involve constant refueling, maintenance, and noise.

One of the first solar-powered water-pump projects was a system installed in Mead, Nebraska, in 1978. This was a U.S. government-sponsored project. The system was designed to irrigate grain fields, using a 25-hp pump, and to run assorted equipment for the drying and storage of the grain produced. The project was successful, demonstrating that photovoltaics makes sense in agricultural applications.

What is especially attractive about solar-powered pump systems is their simplicity. Many of the systems now being installed do not even use batteries. Water storage in tanks has generally been found to be more efficient than electrical storage. In cases of on-demand pumping, however, batteries are necessary because sunlight is available only for a certain number of hours every day.

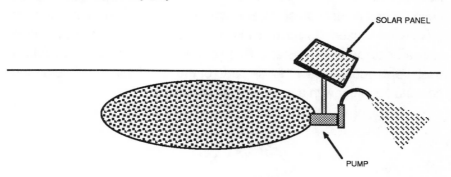

Figure 56: Pond/Stream Pumping System

Solar-powered water pumps are suitable for almost all groundwater applications. For stock watering or flood irrigation, where flow rate is not important, a simple direct-connected system is easily installed. In this case the photovoltaic output is directly connected to the pump, and the water retrieved from the well is stored in a tank or applied directly to the fields for irrigation. Direct-connected pump systems can also be used with applications that are pressure dependent, like drip irrigation.

Solar-powered 12-volt DC pumps are capable of pumping water up to 450 feet high (or 200 psi) or up to 200 gallons per hour. These pumps typically need a 2- to 20-amp solar electric system, depending on the depth of the well and the amount of water to be pumped.

In deep-well applications, motor efficiency is of course critical. Most AC motors, especially those used with conventional deep-well pumps, are traditionally inefficient. In most cases it will, therefore, not even be practical to use an inverter to convert to DC solar electric power for this application. A good pump motor is essential. Then sunlight can work its magic.

Obviously, this fuel-free and quiet equipment can prove an asset for many farms and ranches. Beyond that, if you want to run your pool or spa pump on free energy, solar electric panels now make this possible.

Photovoltaic pumps are a sensible investment. Systems with minimal flow requirements cost around $100, whereas deep-well systems can run up to several thousand dollars. But usually the hotter and sunnier it is the more water is needed. And the sunnier it is the more water can be pumped with a solar-powered system. Therefore, solar electricity is particularly useful and economical in those parts of the world that combine a high demand for water with a great amount of sunlight.

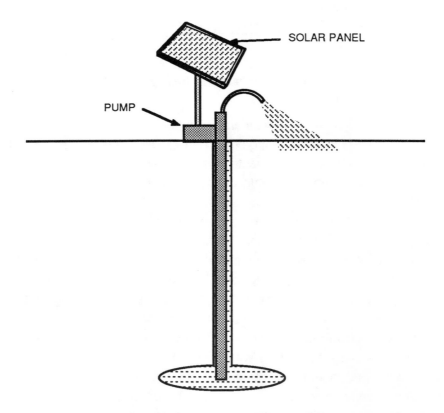

Figure 57: Deep-Well Pumping System

Sunlight is also all that is needed for the installation of a water fountain. This is a simple yet striking demonstration of solar electricity. A solar fountain can be built by using a 2-amp solar panel connected to a small but efficient DC pump. The solar module is directly connected to a 12-volt water pump. So long as there is sunlight, the water will be recycled. The force of flow of the water is governed by the time of day and the available sunlight. In the early morning, the fountain comes to life with a small trickle. By midday, you can enjoy the fountain in full flow. As twilight approaches, the cycle returns to a mere trickle, ceasing altogether when the sun sinks below the horizon. How would you like a sparkling fountain in your garden?

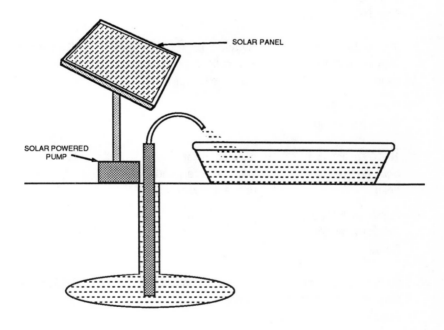

Figure 58: Fountain System

Table 4: Pumps and Their Power Requirements

Characteristics	Application	Power Needs
2-5 gallons per minute, 4-15 feet of head Pump Cost: $100-$200 Brand Names: March, Laing	solar water heating, wood water heating system recirculation, or other very low flow needs	0.75-3 amps
2-3 gallons per minute, 100-200 feet of head Pump Cost: $125-$150 Brand Name: Shurflo	pressurizing water systems, pumping water out of a creek, etc.	2-6 amps
2-10 gallons per minute, up to 40 feet of head Pump Cost: $100-$200 Brand Name: Flojet	RV and boat applications, pumping water up to a tank	2.5-6 amps
10-166 gallons per hour, up to 450 feet of head Pump Cost: $500 Brand Name: Slow Pump	drawing water from shallow wells, ponds, or rivers, and slowly pushing it to a higher location	2-16 amps
90-400 gallons per hour, up to 450 feet of head Pump Cost: $1,000-$3,000 Brand Names: AY McDonald, SolarJack, Tri-Solar	deep-well pumping system	20-100 amps

CHAPTER 10

YOUR HOME AS YOUR SOLAR-POWERED CASTLE

Rumor has it that solar energy is not cost-effective on a large scale. Wrong! The main reason that solar electricity is not in widespread use is simply lack of public education about this form of energy. Solar electricity *is* practical and efficient, whether you own or rent the house you live in, or whether it is a sprawling mansion or a mountain cabin.

As of 1986, there were over 20,000 solar-powered homes in existence, and more are being built every day. Many more should be under construction because those who own solarized homes know that this source of energy makes implicit sense. Not only will they have electricity when the next oil crisis forces power stations to shut down or cut back again, but by choosing solar energy they also help conserve the earth's precious natural resources.

In San Diego, California, there is a new suburban community of 112 townhomes that are solar powered. Each home in this Laguna Del Mar community has two electricity meters. One measures the electricity consumed, and the other keeps track of the power *sold* to San Diego Gas and Electric Company. No storage batteries are involved. Instead, there is a direct tie-in with the local utility company. This is an interesting concept that gives home owners a large measure of independence, while still giving them the means to take advantage of existing utilities. This kind of cooperation between individuals and utility companies is a promising beginning that should prove advantageous to both parties and to the economy at large.

Another solar pioneering project similar to Laguna Del Mar is Solar I in Phoenix, Arizona. In this suburban community of twenty-four homes, the power requirements are met by a central solar electric station, designed to generate 350,000 kilowatt-hours of electricity per year. Each home has its own meter, and the overall idea is that in the course of a year, the production and consumption of energy balance out, so that none of the twenty-four homeowners has to pay any electricity bill at all. The construction cost for this solar power station was added to the purchase price of the homes.

The oldest solar electric settlement in existence is the Papago village of Sohuchuli in Arizona, which has been depending on the sun for all its electricity needs since 1978. The villagers generate around 3,500 watt-hours daily, which takes care of all their basic power requirements.

Figure 59: Solar-Powered Homes in Southern California

That solar power is practical is further demonstrated, for instance, by the Carrisa Plains plant near San Luis Obispo, California. This station, which was the first of its kind, converts sunlight into over 12 million kilowatt-hours of electricity per year, catering to the power needs of over 2,300 homes. Today several other mini power stations using solar cells are performing reliably and economically.

Solar electric homes are a responsible way of living at a time when conventional energy consumption is rapidly depleting and polluting the earth. In 1977, the then Department of Health, Education and Welfare funded a survey that explored young people's attitudes toward different forms of energy production. No fewer than 81 percent thought that the most likely cause of war in the future would be rising energy consumption and decreasing energy supplies. Those young people are now in their forties, and over the past decade they have witnessed a growing conflict over the world's natural resources, particularly oil.

Figure 60: Solar Electric Power Plant near San Luis Obispo

Today the Middle East, which produces about 55 percent of the world's oil, is more unstable than ever, so it is only realistic to look ahead at the economic and political prospects of the 1990s. Most predictions range from moderately difficult to grim. Only irresponsible politicians speak of the immediate future with optimism. In view of this, it is desirable in all respects to prepare for the remaining years of the twentieth century by making responsible choices. Converting your home to solar electricity is definitely one such choice. It is the energy-conscious choice.

Apart from being the ecologically responsible thing to do, solar conversion is economical, if it is done the right way. And it is economical not only for country homes but also city dwellings.

What is more, you won't have to give up any of the amenities to which you have become accustomed. Solar electricity can power any appliance—from a calculator to a washing machine. What you will have to take into account, though, is the fact that it is not efficient to convert electricity into heat. Water heaters and stoves use up massive amounts of energy, which are better supplied by natural fuels like wood or gas. But even then, if you install conventional solar heating panels, you can use sunlight to meet your hot water and heating needs.

The installation itself is simple enough and can be done by any handy person who puts his or her mind to it. In most people's minds, "going solar" means to trap the sun's rays through elaborate systems in order to generate heat. The sun's rays would, typically, fall upon an expensively converted roof where either the air or water would be heated and then circulated around the house. Unfortunately, some consumers, who have had this kind of solar heating system installed, were burned by racketeers who did incompetent installations at high cost. Solar electricity involves none of the elaborate plumbing of solar water heating. We have come a long way since the ancient Greeks and Romans who, facing their own energy problems, first introduced solar architecture.

In using solar electricity, you have the advantage of converting your home step by step, adding more panels and/or batteries as your energy requirements expand and to suit your income. Also, the solar electric system is easy and safe to use. The solar panels have no mechanical parts and so can be expected to work without fail for decades. Neither wind nor freezing cold can interrupt the steady conversion of light into electrical current. There are efficient solar-powered homes even in Alaska where the long winters yield little sunlight.

The best thing of all is that a solar-converted home is an excellent financial investment. You reduce your electricity bills right away and, if you install a whole system, you can even buy your independence from the utility company. At the same time, you improve your home, which is certain to rise in value, especially when the energy crisis rears its head again.

In remote areas, where you would have to pay $4.00 per foot, and more, to have a power line brought to your home, solar electricity is clearly the most economical solution. But if you live in a city and your house or apartment receives direct sunlight, you too can select an appropriate solar electric system. You will have the advantage of pollution-free, gratis energy, while still enjoying the same amenities as your neighbors.

Many of the presently existing solar-powered homes run 80-100 percent on 12-volt DC power. Although it is feasible to power a solarized home using 120-volt AC power, the most cost-effective way is to use low voltage—ideally 12-volt DC, using a 120-volt AC inverter whenever needed or desired. There are three reasons for this choice:

1. Low-voltage appliances tend to be more energy efficient than their 120-volt AC counterparts.
2. Storage batteries are generally 12-volt.
3. Many electrical appliances actually utilize low DC voltage and would only waste energy by going through a transformer that converts 120-volt AC down to 12-volt DC or less.

An important side benefit of the 12-volt DC system is *safety*. There is less shock hazard, which is a blessing when you have young children, who love to explore precisely where they are not supposed to.

Thanks to the phenomenal growth of the RV and marine industries, you can choose from an amazing number of 12-volt appliances.

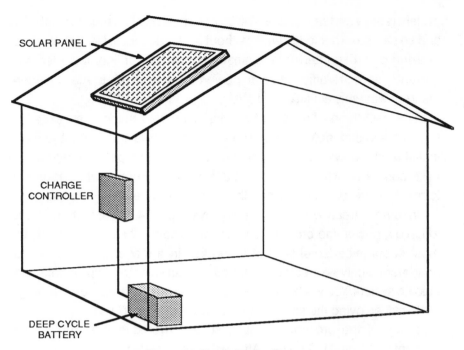

SOLAR PANEL

CHARGE
CONTROLLER

DEEP CYCLE
BATTERY

Figure 61: Home Power System

Power Systems for the Home

The best thing about 12-volt solar electricity is that you can start generating real power for your home for about $500. Below is a starter system that will provide enough electricity for a couple of 12-volt lights, a small television, and other small appliances, such as a tape player or a fan. Please note that in all of the following examples the costs given are approximate and based on 1987 average prices.

1 2-amp solar electric panel	c. $350
1 deep-cycle 105-amp-hour battery	c. $100
1 charge controller	c. $ 75

This system would suit a small cabin, or provide backup power or partial power for a conventional larger home.

Additionally you may want to acquire the following appliances and accessories:

1 12-volt fuse box	c. $ 30
2 12-volt fluorescent lights	c. $ 70
1 12-volt ceiling fan	c. $100-$200
1 12-volt black-and-white television	c. $100

If, for the time being, you decide to go with your present high-voltage appliances, you will need a 500-watt inverter instead, which can be purchased for around $500. But if you are the kind of person who likes to look ahead, I can only encourage you to start switching over to 12-volt appliances right away.

If your present home runs on AC power and you want more or perhaps even complete independence from your local utility company, you have these five options: *emergency power, partial replacement, separate system, utility intertie,* and *complete replacement.* Let me explain in more detail what comprises each option.

Option 1—emergency power: If you plan to continue to use the electric grid most of the time but would like to have some way of running essential equipment during power outages, a modestly priced system like the one

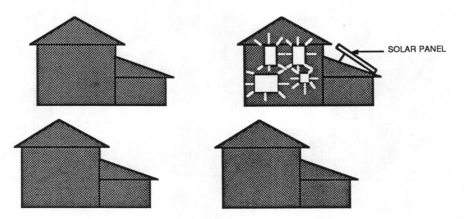

Figure 62: Solar Electricity Providing Emergency Power During Blackouts

described above is the answer. It will supply sufficient electricity that you won't have to sit in a dark or overheated room, can stay in touch with the news either through radio or television, and can keep your tropical fish alive by running the aquarium pump and heater, if necessary.

A slightly larger system can produce enough emergency power for a refrigerator and you won't ever have to argue with your insurance company over exactly how much food got spoiled during the last blackout.

Option 2—partial replacement: If you cannot afford to make your home completely independent from the power grid right now, but want to break away from it slowly, you can rewire a room or two to run off your own solar electric system. In this way you can still benefit financially—and psychologically—and your system can also grow with you, as your need or desire for electrical independence grows. This is also a marvelous way of demonstrating the practicality and efficiency of photovoltaics to your friends and neighbors. Your own good sense and courage could be the seed for the solarization of your entire neighborhood.

SOLAR PANEL

Figure 63: Partial Replacement of Utility Supply with Solar Electric System

SOLAR PANEL

Figure 64: Solar-Powered Malibu Lights

Option 3—separate system: You may want to stay on the utility grid, but use solar electricity for some of your appliances—just for the fun of it and also to experiment a little with photovoltaics. Malibu or garden lights, for instance, are an ideal candidate for solar power. Generally, these lights are designed for 12-volt DC power and are therefore sold with a transformer that, rather inefficiently, changes 120 volts to 12 volts. It is a lot more efficient to run Malibu lights directly off a solar-powered 12-volt battery system.

You may think of other similar applications or needs, perhaps in your garage, greenhouse, shed, or workshop.

Option 4—utility intertie: Are you ready to make a major commitment to solar power and the independence it will afford you, but still want 120-volt power and grid connection? If so, you don't need to invest in storage batteries. You can continue to utilize your present house wiring. The best part of this method is that, under current utility laws, your utility company is required to *buy* the excess power you generate. What this means is that

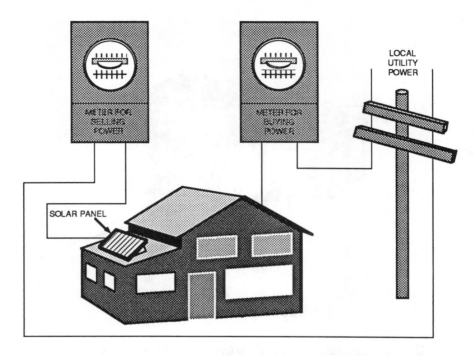

Figure 65: Utility Grid Connection—Solar Electric System

during the day you sell excess electricity to the company and at night you buy it back. Most utility companies must buy your power at the rate it would cost them to produce it. At night the billing rate is usually the lowest, since it falls outside the hours of "peak usage."

Now, if you size your system correctly, your daytime power selling can equal your nighttime cost. An ideal "balance of power," wouldn't you say? You will still receive your monthly or quarterly bills, but you won't owe a penny. You may even be able to generate more power than you use, thus actually earning money from the utility company. Wouldn't that be a pleasant switch: to receive a check with your bill rather than to have to write one every month or quarter?

Option 5—complete replacement: Total energy self-sufficiency, or happy independence from the power grid, is the most expensive of our five options, but it is also the most rewarding. In cases where conventional power is not available, it is still economical. Moreover, if you design and build your home *correctly,* complete electrical independence will be both efficient and economical, no matter where your house is located. I will be looking at this option in detail in the next chapter.

The 3.5-kilowatt photovoltaic system that was installed in the Indian village of Sohuchuli was designed to power water pumps, lights for fifteen houses, a communal washing machine and sewing machine and, not least, fifteen 4-cubic-foot refrigerators for ninety-two people. This pioneering project proved successful and served as an example to foreign countries to experiment with solar power in their generally unelectrified rural districts.

If an entire village can become power-independent, so can you—and, of course, at a much reduced cost. Then again, you may want to consider joining forces with your friends or neighbors and cooperate on a solar power project that could create not only independence for each family but also a new spirit of adventure and sharing.

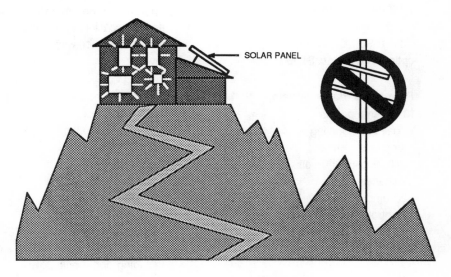

Figure 66: Complete Grid Independence

Table 5: Home Power Systems

System	Daily Output in Watts	Equipment
Basic Cabin System	150-200	2 lights — 1 hour
1 2-amp panel		1 TV — 1 hour
1 pair mounting brackets		1 radio — 1 hour
1 105-amp-hour battery		
1 charge controller		Capable of meeting
Cost: c. $500		minimal power needs
Medium Cabin System	1,000	3 lights — 4 hours
4 2-amp panels		1 TV — 3 hours
4 pairs of mounting		1 radio — 1 hour
brackets		1 fan — 1 hour
2 220-amp-hour 6-volt		
batteries		Capable of meeting
1 10-amp charge		medium power needs, of
controller		providing partial or back-
1 1200-watt inverter		up power for conventional house,
(optional)		and of occasionally running
Cost: c. $2,000		an AC appliance with an
optional inverter: c. $1,300		inverter
Power Plus Starter System	1,500	5 lights — 4 hours
13 1.5-amp panels		1 TV — 3 hours
13 pairs of mounting		1 stereo — 2 hours
brackets		1 fan — 2 hours
4 240-amp-hour 6-volt		1 computer — 1 hour
batteries		1 blender — 5 mins.
1 20-amp charge		1 vacuum — 30 mins.
controller		
1 1,500-watt inverter		Capable of providing all the 12-volt
Cost: c. $5,000		power for the basic needs of a cabin,
		including occasional power
		for AC appliances

Independent Power System   6,000   Can meet the electrical
52 1.5-amp panels   needs of a home compar-
or   able to a conventional
24 3-amp panels   AC-powered home
52 or 24 pairs of
   mounting brackets
12 1,015-amp-hour 2-volt
   batteries
1 charge controller
1 24-volt 2,500-watt
   inverter
Cost: c. $20,000

CHAPTER 11

POWER TO THE NEW HOME—DO IT RIGHT!

This chapter is for those who are designing a new home or cabin from the ground up and want to make it completely energy independent. But many of the hints given here will also be useful to others who are adapting their existing home, office, or cottage to solar electricity.

As an ancient proverb says, "Home is where the heart is." A home is so much more than a house. It is a place where you can be at ease and feel secure, where you can love freely and are loved. Modern architecture has almost forgotten how to make homes. To create a home out of a house is a great art, and it is an art that must come into play from the moment a house is conceived.

So, if you want to make yourself a home, rather than build a house, you must take into account all the many factors that add up to ease, comfort, safety, and human warmth. I feel that solar electricity is definitely an integral part of such home making. Therefore it should be planned right from the start.

Heating and Hot Water

A *cold* home is almost a contradiction in terms. Traditionally, to be home meant to be in front of a friendly hearth. As Emerson wrote in his *Journals:* "My idea of a home is a house in which each member of the family can on the instant kindle a fire." While it is fun to watch the flickering flames, really it is the fire's warmth that makes you feel at home. And what a homey luxury a hot shower or bath is! So, when planning your new home be sure to get the heating and hot water supply right. Here are some important points to bear in mind:

1. Take care of your heating loads through a combination of solar and alternative energy means, but make maximum use of passive solar design principles. There are five main principles to consider in designing your home:

a) Insulate and weatherstrip thoroughly and carefully.
b) The long axis of the building should be oriented toward the south. (All references to "south" are valid only for the northern hemisphere and should be replaced with "north" for homes located in the southern hemisphere.)
c) The majority of the windows should be placed on the south side of the building.
d) An appropriate overhang should be built on the south side of the building.
e) If cooling is a concern, the roof should be covered in a light color.

In most cases, these five principles insure that 50 to 75 percent of your heating and cooling needs are met. There are a number of excellent books on this subject currently available, which you can find in any good bookstore or your local library. Some are listed in this book under Recommended Reading.

2. Incorporate a solar water-heating system. A conventional flat-plate collector or, if you prefer, a passive solar water heater is the most trouble free and inexpensive water-heating system available.

3. Provide adequate backup for the solar water heater, through either a wood or propane water heater. With a tankless water heater, no gas is burned to keep the water hot when you are not using it. No energy is wasted through heat loss from the tank or water line. When you need hot water, you simply turn on the tap and the tankless water heater supplies it instantly.

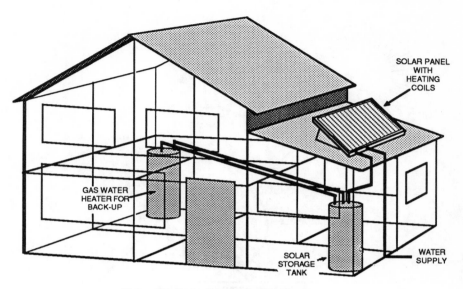

Figure 67: Solar Water Heating System

An efficient airtight wood stove can provide backup heating for really cold winter days when the sun peeks only occasionally through the clouds. If, in addition, you place a stainless-steel water jacket in your stove, you have a reliable backup source for hot water.

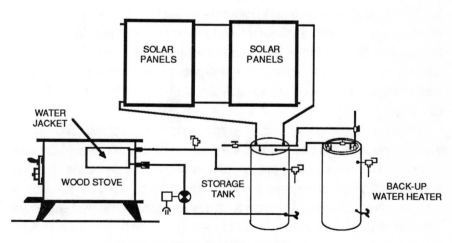

Figure 68: Solar/Wood Connection

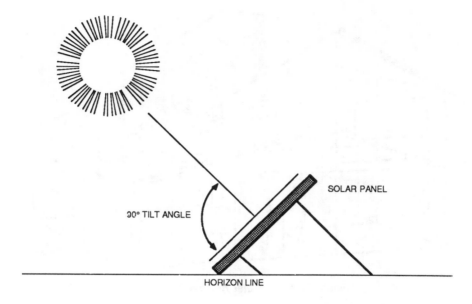

Figure 69: Solar Electric Panel Location

Locating the Solar Panels

In order to generate as much power as possible, the solar panels need maximum exposure to sunlight for the longest possible time. In the United States (and the rest of the northern hemisphere) the panels should face directly south. In the southern hemisphere they should face north. Depending on your location and local climate, you may want to modify this rule slightly.

For example, if the mornings are foggy where you are, you can position your panels to face a little more to the west, since the sun rises in the east and sets in the west. If there are large trees or other obstructions blocking the sun on the eastern exposure, you can have the panels face more to the west. Study the course of the sun for a day to discover the best exposure position. This is a wonderful way of getting in touch with your natural environment.

Tilt Angle

For optimum sun exposure, your solar panels should be installed at an angle perpendicular to the sun at noon. In countries closer to the equator the angle should approach 90 degrees, while in Alaska, for instance, it should approach 30 degrees.

As the earth moves around the sun, thereby creating the annual seasons, the angle of incoming sunlight changes over the year. You may, therefore, want to consider adjusting your solar panels two, three, or even four times a year for maximum efficiency. Most people, however, mount the panels permanently at an angle close to the winter position perpendicular to the sun at noon. This is the easiest and simplest way. During the summer the loss of surface exposure is compensated for by the longer hours of exposure.

Fortunately, the winter angle is close to the roof slope in most cases. For economy and convenience, the panels can be mounted directly on the roof, providing it slopes in the right direction.

Maintenance of Solar Panels

Once your solar panels are installed, the good news is that you don't have to think about them again—other than enjoying the fact that they provide you with free electricity. In other words, there is virtually no maintenance. If you live in a place where there is a lot of dust during the summer and little wind or rain to wash it off your roof, you may occasionally want to hose down the panels. In most cases, however, Mother Nature will do all the maintenance that is necessary.

How Many Panels Do I Need?—Sizing Your System

In order to size your solar panels correctly, you must first determine the electricity needs of your home. You do this by listing all the appliances you want to run and estimating how many hours they are likely to be used on an average day. Then you multiply the number of hours by the amperage consumed. When you divide the total amp-hours by the number of available hours of sunlight per day, you obtain a figure that will give you the right panel size. Let me walk you through this calculation step by step.

Step 1: List all the appliances you plan to use, together with their respective amperage consumption. The amperage figure is usually given on an appliance or in the instructions for it. If it is given in watts, all you have to do is divide the figure by the voltage to get the amperage.

Since low-voltage appliances are the most efficient in connection with solar electric systems, here is a list of common appliances and their power consumption in amps:

Lights

8-watt fluorescent	0.7
15-watt fluorescent	1.4
30-watt fluorescent	1.9
25-watt incandescent	2.1
21-watt quartz halogen	1.8
30-watt halide	3.0

Fans

6-inch blower	0.25
12-inch exhaust	2.0
36-inch ceiling	1.2
52-inch ceiling	2.5

Broadcasting Appliances

AM/FM car radio with tape player	0.5
5-inch black-and-white television	0.6
12-inch black-and-white television	1.4
13-inch color television	4.1
satellite receiving system	0.6-1.5

Kitchen Appliances

blender	6.0
coffeepot	11.0
toaster	20.0
small portable refrigerator	2.0
standard refrigerator	2.0-6.0

Water Pumps

small pump (2 gallons/min.)	0.5
shallow-well pump (depending on flow rate and head)	1.0-8.0
deep-well pump (depending on flow rate and head)	5.0-20.0

Tools

drill	12.0
circular saw	20.0
chain saw	100.0

Miscellaneous Appliances

clock	0.1-0.3
CB radio	0.5
radio/telephone (depending on transmission power)	0.3-12.0
curling iron	3.5
vacuum cleaner	9.0
clothes iron	10.0
grain mill	15.0

Step 2: Multiply the amperage consumption of each appliance by the average number of hours each appliance will run. This will give you the total number of amp-hours.

Table 6: Calculating Total Amp-Hours (Example)

Load Consumption (in Amps)	Average Number of Hours Used Per Day	Amp-Hours Per Day
2 20-watt fluorescent lights for living room (1.7 x 2 = 3.4)	x 4	= 13.6
1 20-watt fluorescent light for the kitchen (1.7)	x 2	= 3.4
1 15-watt fluorescent reading light for bedroom (1.4)	x 1	= 1.4
1 15-watt fluorescent bathroom light (1.4)	x 1	= 1.4
1 AM/FM radio (0.5)	x 2	= 1.0
1 12-inch black-and-white TV (1.4)	x 3	= 4.2

Step 3: Total up the number of amp-hours: 25.0

Step 4: Allow a 20 percent energy loss for your battery's storage efficiency. In other words, multiply the total amp-hours by 1.2. In our example, this would give you a grand total of 30 amp-hours.

If you are using an inverter (see the next section), you should factor in an additional 15 percent energy loss. That is, multiply the total amp-hours, including the 20 percent for battery loss, by 1.15. In our example, you would thus have to add another 4.5 amp-hours to the grand total of 30 amp-hours.

Step 5: In order to obtain the size in amps of the solar panels needed for your level of consumption, you now divide this total (30 amp-hours) by the number of available hours of sunlight per day, calculated on the basis of average local weather data or using a chart like the one given on page 13 for the United States. Thus:

$$\frac{30 \text{ amp-hours}}{5 \text{ hours}} = 6 \text{ amps}$$

Step 6: Choose large enough panels to provide this number of amps. You could, for instance, install three panels of 2 amps each, or two panels of 3 amps each. A 3-amp panel would have dimensions about 1′ x 4′ and cost around $400.

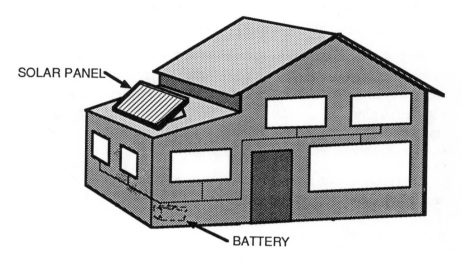

SOLAR PANEL

BATTERY

Figure 70: A 12-Volt House

Converting DC into AC Current

If you have installed a 12-volt DC system but some of your appliances are still running on 120-volt AC current, you don't need to dispose of them, though you might want to consider replacing them in the long run. In order to run your 120-volt juicer or vacuum cleaner on a 12-volt battery system, you will need an inverter.

Inverters are typically rated in watts, and you have to install one that will meet your requirements. Sizing your inverter is similar to sizing your solar panels, except that it is easier to calculate it in watts rather than amps, since appliances and inverters are rated in watts.

Step 1: List all the AC appliances that you wish to run on your 12-volt system.

Step 2: List the wattage for each appliance in a separate column. Where only amperage is given, you can convert this into wattage by using the simple formula: W = V x A (see Chapter 5).

Step 3: Estimate the maximum wattage you are likely to use at any one time. This gives you the continuous output power.

Step 4: List the *surge rating* for each appliance powered by an AC motor. Most AC motors use more power when they are started up. In some cases this is four to five times more than the power requirement during continuous running. For instance, an AC induction motor of 1 hp requires a start-up wattage of 10,000 watts, but runs efficiently on 1,800 watts. For smaller appliances the surge rating is smaller.

Step 5: Choose an inverter that meets your total wattage requirements and has a high enough surge rating to handle your start-up loads.

Batteries and Backup Power

I have dealt with batteries at length in Chapter 6. Here I simply want to reemphasize that the best battery for your solar-powered home is the deep-cycle battery. It is efficient and affordable. If your funds permit, however, it may be worth acquiring a more high-powered industrial battery.

Depending on the climate, you may also want to consider installing a backup generator, which can keep your batteries charged during pro-longed periods of inadequate sunlight or provide you with extra power when you need it. Propane generators have been found to be efficient for this purpose.

When buying your generator, make sure that its output meets your wattage requirements.

Wiring for Power

Solar panels and batteries of course need to be wired up. *But* you have to use the right kind of wire, and also connect it up in the right way, to achieve optimal efficiency. And that's what this section is about.

Wires come in a variety of sizes. Many wires are made of copper encased in plastic. The wire is either solid or stranded, and the latter kind is the best. A third type of wire is known as Romex, which consists of two or three separate solid wires covered in plastic. It is commonly used with AC current.

Low-voltage systems call for wires that are larger than the typical 120-volt AC wires, which are mostly 14 gauge. The reason for this is, as I have explained earlier in the book, that each wire—like a water hose—has built-in resistance. A 12-volt current would encounter too much resistance in a small-gauge wire, causing it to overheat. This would mean a loss of voltage and inefficient operation of the appliances connected to the wire.

Heavier-gauge wires help prevent voltage drops. You can check the efficiency of your present wiring right now: switch on your television and a couple of lights, and then start your popcorn maker or another appliance on the same circuit. If the lights grow a little dimmer and your television picture jumps, there has been a voltage drop.

Thus, #14 wire is commonly used in 120-volt systems, but this would not do in 12-volt systems, *unless* the wattage of the appliance is very low and the distance between power source and load is short. Even for 120-volt systems, it is preferable to use #12 wire, even though it is more expensive. It is safer, and #10 wire allows you to switch over to a 12-volt system later, without having to rewire the whole house. Heavier #8 wire is absolutely necessary for your car to handle the high amperage of the battery. To connect your solar storage batteries together, #4 wire is recommended.

The longer a wire is the greater the resistance will be and consequently the greater the loss in voltage. For instance, if you run a 12-volt current through 100 feet of #10 wire, you will get 11 volts at the other end of the wire. Using #8 wire, the loss is only 0.6 volt, whereas #0 wire has a loss of only 0.1 volt.

The chart below gives you the correct wire gauge (from #12 to #0) for a given amperage and distance (or length of wire).

The following figures represent the maximum distance from power source to load for copper two conductor wire in 12-volt systems.

Wire

Amps	12	10	8	6	4	2	0
2	85	130	220	360	560	900	1500
5	35	56	90	114	225	362	600
10	18	29	45	77	112	181	300
15	11	18	30	47	75	120	200
20	8.5	13	22	36	56	90	150
25	6	11	17	29	45	72	120
30	5	8	15	25	37	60	100
50	3	5	8	15	22	36	60

Figure 71: Wiring Chart

In a DC system, the current flows in one direction only. This means that you must run *two* wires to each plug, outlet, switch, or appliance—one wire for positive, the other for negative. It is important to know which wire is positive and which is negative because if you cross wires, you can severely damage your wiring and appliances. Therefore the safest installation includes a 12-volt fuse panel, similar to the one in your car's electrical system. Since 12-volt fuses tend to be hard to read, some of these fuse distribution panels incorporate small test lights.

Twelve-volt DC appliances should never be plugged into 120-volt AC outlets and vice versa. If you are using both types of current, you should install separate outlets for each. The standard plug and outlet for a 12-volt system are those used for cigarette lighters. In this type of connection, the outside rim carries the negative current (and also serves as ground), whereas the pin is positive.

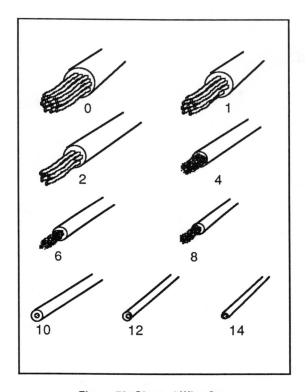

Figure 72: Chart of Wire Gauges

There are attractive 12-volt switches, plugs, and outlets available in a wide range of single-purpose and combination styles. These have higher amp ratings than their 120-volt counterparts, which should be clearly marked so that they cannot be confused with each other.

The *Conserve Switch* is a new idea in low-voltage electronics. This 12-volt rheostat allows you to vary the speed of low-voltage motors and to dim incandescent or even fluorescent lights. Unlike conventional rheostats, the *Conserve Switch* wastes very little power when regulating appliances. It is therefore ideal for solar electric systems where power conservation is key.

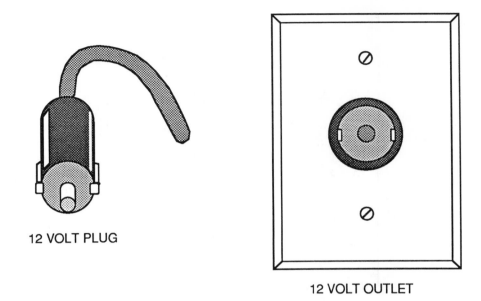

12 VOLT PLUG

12 VOLT OUTLET

Figure 73: Plug and Outlet for 12-Volt System

CHAPTER 12

COMMON HOUSEHOLD APPLIANCES— A SAVER'S GUIDE

Lighting

We've come a long way since Edison opened the world's first commercial power plant in 1882. Around the globe millions of light bulbs and fluorescent tubes light up the night sky over cities and villages. Unless we happen to live in the outback, where a noisy and not always reliable generator supplies our electric needs, we don't even think about lighting anymore—that is, until there is a sudden power failure or a bulb blinks out on us.

In a typical modern home, lights are fitted just about anywhere—from ceilings and walls to stair rails and furniture. They are the most visible reminder of our conspicuous energy consumption. So, conservation and saving can begin right there. Even though it is true that lighting is not the biggest area of energy consumption, it is nonetheless appropriate to consider how power can be saved with the right lighting system.

There is a widespread misunderstanding that the higher the wattage of a light bulb the more light it yields. In fact, however, there isn't a one-to-one relationship between wattage and luminosity. To find out how brightly a bulb will shine, you need to look up its number of *lumens,* as given on the package. Thus, a 60-watt bulb may have 855 lumens. Or a three-way bulb may have 50, 100, and 150 watts at 580, 1640, and 2220 lumens respectively.

Do we need to have such high luminosity? The answer is No in most cases. If you were to illumine only the immediate area involved in carrying out a task that requires you to see well, all you would need is 30-40 lumens. You would be able to read a book or newspaper without eye strain. If a somewhat larger area needs to be illuminated, 50-70 lumens would be adequate in most cases. For general indoor recreation it could be as low as 10 lumens. For tasks involving more visual concentration, such as sewing, you would want to have 100-200 lumens. What these figures tell us is that our normal lighting is a luxury costing us money.

There are three principal kinds of lights—incandescent, high-intensity discharge, and fluorescent. The ordinary light bulb is incandescent. It is the oldest type in existence and also the most uneconomical. High-intensity discharge lights yield up to five times more light than incandescent bulbs, while consuming the same wattage and lasting up to thirty times longer.

The most efficient form of lighting is fluorescent. It uses a quarter of the power of incandescent light. What this means is that a 25-watt fluorescent light will, roughly speaking, be as efficient as a 100-watt incandescent light. There is another bonus: a fluorescent tube will last about twenty times longer than an incandescent light bulb. True, fluorescent light doesn't have the same warm quality that incandescent light has, but there are now fluorescent tubes that are color-corrected.

There are extra savings in switching from 120-volt AC to 12-volt DC current. For instance, a 12-volt DC incandescent light is roughly twice as efficient as its 120-volt AC counterpart. That is to say, using a 25-watt bulb at 12 volts DC, you get the same amount of lumens as you would using a 50-watt bulb at 120 volts AC. It is one of the physical characteristics of incandescent lighting that it produces more lumens per watt if the amperage is increased. And in low-voltage systems, the amperage is naturally higher.

Twelve-volt incandescent bulbs are available to fit 120-volt fixtures, though you may have to upgrade the wiring.

The most efficient form of lighting is undoubtedly 12-volt DC fluorescent lighting. Thanks to the RV industry and a growing "independent

Figure 74: Types of 12-Volt Lighting

home" network, 12-volt DC fluorescent lights are becoming easily available. A 16-watt fluorescent tube running on 12 volts DC will yield as much luminosity as an 80-watt incandescent bulb. What is more, the flicker that is common to fluorescent tubes running on 120 volts AC is usually not found in 12-volt DC versions. The reason for this is that 12-volt DC tubes operate at very high frequencies of 30,000 cycles or more, whereas 120-volt AC tubes operate at 60 cycles, which has been found tiring to the eye muscles. This side effect is absent in 12-volt DC fluorescent tubes, which give out a steady and cool light.

They are available in many different shapes, configurations, and styles, suitable for every imaginable application. Altogether, 12-volt DC fluorescent lighting is the primary candidate for any independent power system.

Heating Appliances

Low-voltage electricity is inefficient as a means of generating heat. Therefore, where heat is needed—to keep the living area warm and for cooking or hot water—another energy source is required. A solar collector on the roof, if the system is designed and installed properly, might be your first choice. Or you may consider an airtight wood-burning stove. This type of stove sucks in just enough air to allow slow combustion, but does not, as do most conventional fireplaces, drain the house of heat. While a good wood-burning stove costs more, it will also save you dollars in wood, because it burns more slowly and efficiently.

Finally, you may want to consider propane for backup heating. Catalytic heaters are fuel efficient and are nowadays available for 12-volt systems.

Refrigeration

Cooling, like heating, is high in energy consumption. Therefore, with the exception of ventilation, it is best handled by non-solar means. Propane refrigerators do an efficient and economic job. They require little maintenance and work reliably for many years.

If this isn't an option for you, there are 12-volt DC refrigerators, but the most energy-efficient designs tend to be expensive because they are manufactured only in limited quantities. The best custom-made 12-volt DC refrigerator available today is *Sunfrost*. It is extremely energy efficient, though it does cost over $1,000. At any rate, you should carefully check out the power requirement of a refrigerator before you spend money on it and see whether your solar system can cope with what is likely to be a heavy load.

Your refrigerator's power consumption is dependent on its size in cubic feet, whether the unit has a freezer, whether the freezer is separate from the refrigerator, what temperature settings are used for refrigerator and freezer, and how regularly you defrost the unit. Here are some dollar-saving tips:

1. Get the right size unit for your needs. A two-person family generally finds 8 cubic feet adequate. Add 2 cubic feet per extra person.
2. Get a unit in which the freezer is completely separate from the refrigerating section. Self-defrosting units are not energy efficient. The most energy-efficient freezers are the horizontal chest types because when you open the door the cold air tends to drop back into the freezer rather than fall out at you, as is the case with upright freezers.
3. Run the freezer at 0-5 degrees and the refrigerator at 34-37 degrees.
4. Defrost regularly. Don't wait until there is a huge buildup of ice. Allow no more than a ¼-inch layer to accumulate.
5. Keep your freezer and refrigerator filled, but allow enough space for air to circulate.
6. Make sure the freezer and refrigerator doors seal properly.
7. Remove the dust from the condenser coils at the back of the unit at least four times a year.

Washers and Dryers

Laundry equipment is another energy guzzler. Again, you should shop around for the most efficient brands. For economy, your washer should be fitted with controls for water temperature and water level. An additional useful feature is a water suds saver. Heating up the water for your washing machine represents a big hidden cost, and it pays to carefully regulate the water level for each wash and to determine whether cold or warm water won't suffice for a given load. The older single-speed tub washers are easily retrofitted with an efficient 12-volt DC motor.

Dryers are power hungry whichever way you look at it. But once you have purchased a dryer that looks the most efficient to you, there are a few things that you can do to cut down on its energy consumption. First of all, don't install it, as many people do, in an unheated part of your house, because in cold weather your dryer will do its best to heat up its surroundings as well. Don't overdry your clothes, and use the remaining heat of one load to start up the next one. Finally, regularly clean the lint filter on your dryer (and also on your washer if it has one).

Again, if you want an efficient dryer, be sure it uses gas. In this way, one propane tank can supply all of your heating needs, whether it be for drying clothes, cooking, or heating your water.

A lot of this is just common sense. Of course, the most efficient clothes dryer is a "solar dryer"—the old clothesline, which uses the free energy of wind and sun.

Figure 75: "Solar Drying"

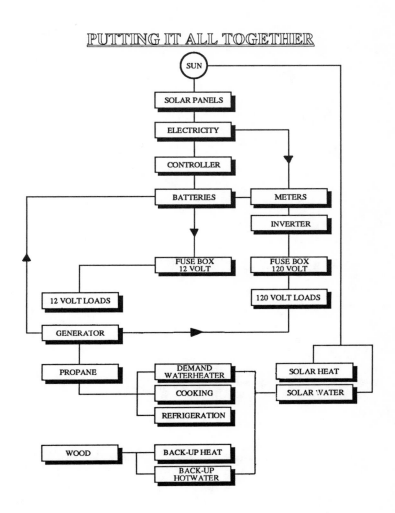

Figure 76: Putting It All Together

AFTERWORD

SEEING THE LARGER PICTURE

We live in dangerous times. Every year, hundreds of billions of dollars are spent by the governments of the world to arm their respective countries. Since the detonation of the first atomic bomb in 1945, the proliferation of nuclear weapons has reached perilous proportions. There are now enough nuclear warheads to dump over 3 tons of TNT on every single person alive.

Moreover, those nations that have the capital to develop nuclear reactors are responsible for all 1,800 metric tons of radioactive waste every year—waste for which there is no safe disposal. The terrible accidents of Three Mile Island and Chernobyl are vivid reminders that these installations are unsafe.

Most importantly, there is an established link between the proliferation of nuclear weapons and nuclear power stations. And yet the current economic situation of the world is putting pressure on governments to continue to invest in the nuclear power industry and to build new reactors. There is a great reluctance on the part of the world's governments to promote alternative energy sources, such as wind and solar power. But as I have explained already, nuclear power is economical only because that industry receives massive government support.

And so nuclear weapons continue to be manufactured, even though the existing supply is more than sufficient to devastate our planet and wipe out the human race and all life.

Moreover, research has shown that, apart from any moral considerations, peace is a better investment than war. The production of nuclear arms is both capital intensive and energy intensive. A society cannot remain, or even be, healthy if it channels a major portion of its funds to build devices whose use would be utterly self-destructive and which, therefore, can never be used.

The Manhattan Project, which ushered in the nuclear age, was a good demonstration that a difficult goal may be accomplished swiftly when a government steps in with single-minded support. Hence Mark Goldes of the AESOP Institute proposed a "Brooklyn Project," which would be an accelerated program to confront the extreme urgency of *reducing* the use of nuclear power, in addition to eliminating the need for imported oil.

Oil has become a major political issue in the world. Despite lower oil prices, the western world is still dangerously dependent on oil from the Middle East, a region which is a time bomb waiting to go off. Clearly, this dependency should be ended as quickly as possible to rebalance the economic and political forces in the world. Little is being done, though, to bring about the necessary changes.

Meantime, the world's population is growing by around 250,000 people every day. The earth's resources are rapidly being depleted. Our environment continues to be polluted, and we are already suffering the first of Nature's return attacks, such as increased ultraviolet radiation, the greenhouse effect, and acid rain.

This leaves you and me to assume responsibility for our lifestyle. We don't have to wait for governments to take action. We can't afford to wait. We *can* make a difference in our immediate environment. Using solar electricity and solar thermal energy is an important step in the right direction.

Massive centralization of electricity production is making us particularly vulnerable: the failure of a single power plant can immobilize a city as readily as the dropping of a nuclear warhead. Prolonged power failure could prove a serious threat to human life, health, and the economy of a city and its dependents.

Photovoltaics, or solar electricity, has the potential of bringing about significant change. Professor Barry Commoner, a leading spokesman for the ecology movement, compares this turning-point to the abolition of slavery. Large-scale implementation of this new technology could:

1. release us from oil's umbilical cord and greatly relieve one area of potential global conflict;
2. provide decentralized, non-polluting electrical power;
3. allow the third world to take a quantum leap toward modernization;
4. slow the proliferation of nuclear weapons by making nuclear power plants obsolete, and
5. help develop an increasingly humane economy.

Today, more than at any other time in history, our lives are all interdependent. Therefore, it is necessary for everyone to see the larger picture. Only then can you and I begin to make responsible decisions. We must become conscious users of energy.

In this book I have shown you many practical ways in which you can make this life-enhancing solar electric vision come true in your own home environment. Now is the time.

Figure 77: The Earth Is Our Home

GLOSSARY

Alternating Current (AC) — The standard electrical current in the United States, in which the flow of electrons is reversed 120 times per second (i.e., 60 cycles per second).

Amorphous — The condition of a solid in which the atoms are not arranged in an orderly pattern, as in amorphous solar cells as opposed to crystalline cells.

Amperage, Amp (A) — A measure of the flow rate of the electrical current, whereby 1 amp is generated by a 1-volt current flowing across a resistance of 1 ohm. It is the passage of 63 billion billion electrons per second.

Ampere-hours (AH) — A current of 1 amp running for one hour.

Battery — A device that can store electrical energy by changing electrical current into chemical energy and releasing it by reversing this process.

Controller — An electrical device that regulates an electrical current in order to prevent overcharging of batteries.

Diode — A device that prevents an electrical current from running backward and thereby causing, for instance, the discharge of a battery.

Direct Current (DC) — The type of one-directional current supplied by batteries and solar electric panels.

Electrical Current — A movement of electrons, electricity.

Grid — The network of transmission and distribution lines and transformers used in utility central power systems.

Inverter — An electrical device that changes DC current (from batteries) to AC (regular household current).

Kilowatt-hour (KWH) — 1,000 watts consumed over one hour, which is the measure used by utility companies in calculating your monthly power consumption.

Line Loss — The voltage drop over a distance of wire when the wire used is too small for the load.

Load — The amount of electrical power being consumed at any given moment by any device or appliance.

Megawatt (MW) — 1 million watts or 1,000 kilowatts.

Module — An assembly of solar cells, a solar electric panel.

Ohm — A measure of resistance to the flow of an electrical current.

Parallel Connection — A method of connecting two or more electrical devices (such as batteries or panels) so that the positive pole is joined to the positive pole and the negative pole to the negative pole. This connection increases the amperage, while the voltage remains constant.

Photon — A particle or quantum of light moving at the speed of 186,300 miles per second.

Photovoltaic (PV) — Referring to the conversion of light (Greek "photos") into electricity.

Series Connection — A method of connecting two or more electrical devices so that opposite poles are joined, which leads to an increase in voltage.

Silicon — A chemical element found in sand and quartz. It is an excellent semiconductor and so is widely used in the production of solar electric cells.

Solar Electric Cell — A photovoltaic cell converting sunlight directly into electricity.

Sulfation — A condition occurring in an unused and uncharged battery in which crystals of lead sulfate form on the plates, destroying the battery.

Voltage, Volt (V) — A measure of the force or pressure given the electrons in an electrical circuit. Thus 1 volt produces 1 amp of current when flowing over a resistance of 1 ohm.

Wattage, Watt (W) — A measure of electrical activity or rate of work equivalent to 1 volt-ampere. Thus 1 amp flowing at 1 volt produces 1 watt of power.

Watt-hour (WH) — The quantity of power used when 1 watt is consumed for a period of one hour.

RECOMMENDED READING

Arvill, Robert. *Man and Environment: Crisis and the Strategy of Choice.* Harmondsworth, England: Penguin Books, repr. 1978.

Bergman, Elihu et al., eds. *American Energy Choices Before the Year 2000.* Lexington, Mass./Toronto: Lexington Books, 1978.

Brown, Lester R. et al., eds. *State of the World 1986.* New York/London: W. W. Norton, 1986.

Callenbach, Ernest. *Ecotopia Emerging.* Berkeley: Banyan Tree Books, 1981.

Clark, Wilson. *Energy for Survival: The Alternative to Extinction.* Garden City, N.Y.: Anchor Books, 1975.

Commoner, Barry. *The Politics of Energy.* New York: Alfred Knopf, 1979.

Daniels, Farrington. *Direct Use of the Sun's Energy.* New York: Ballantine Books, 1974.

Elgin, Duane. *Voluntary Simplicity: An Ecological Lifestyle that Promotes Personal and Social Renewal.* New York: Bantam Books, 1982.

Ewers, William. *How to Use Solar Energy.* Tucson, Ariz.: Sincere Press, 1976.

Freeman, S. David. *Energy: The New Era.* New York: Vintage Books, 1974.

Halacy, D. S., Jr. *The Coming Age of Solar Energy.* New York: Harper & Row, rev. ed. 1973.

Holdren, John and Herrera, Philip. *Energy: A Crisis in Power.* San Francisco/New York: Sierra Club, 1971.

Maycock, Paul D. and Stirewalt, Edward N. *A Guide to the Photovoltaic Revolution.* Emmaus, Pa.: Rodale Press, 1985.

Miller, G. Tyler, Jr. *Energy and Environment: The Four Energy Crises.* Belmont, Calif.: Wadsworth Publishing, 1975.

Rothchild, John. *Stop Burning Your Money: The Only Home Energy Guide You'll Ever Need.* New York: Penguin Books, 1982.

Stobaugh, Robert and Yergin, Daniel, eds. *Energy Future: Report of the Energy Project at the Harvard Business School.* New York: Vintage Books, rev. ed. 1983.

INDEX

THE SOLAR ELECTRIC NEWSLETTER

This new monthly publication
- helps you keep abreast of the latest technological developments in the solar electric field
- informs you about new and established photovoltaic consumer products
- updates you on solar electric stocks and investments
- provides you with useful practical information on solar electric applications and energy-conserving methods and appliances
- brings you news about the growing solar community around the globe
- comments on government and world affairs affecting the solar electric industry and its consumers
- looks ahead to future breakthroughs and possibilities
- lets *you* share your *own* views, news, and experience with solar electricity

The Solar Electric Newsletter is written for all those who share in the adventure of applying this advanced and exciting technology in their lives and who are committed to independent energy-conscious living and a planet that is pollution free.

Yearly subscription is only $25.00. And as a subscriber you are entitled to a 10 percent discount on all solar electric products sold by Solar Electric. Your subscription can pay for itself with your first purchase!

Please fill out the form on the next page in BLOCK LETTERS and send it in an envelope together with your check or money order to:

Solar Electric
175 Cascade Court
Rohnert Park, CA 94928
Tel.: (707) 586-1987

FREE SOLAR CELL

This copy of your *Solar Electric Book* comes with the exceptional offer, by Solar Electric, of a *free* working solar cell worth $5.00. The cell is wired to a musical chip that plays whenever the cell is exposed to light.

This is a practical demonstration of the simplicity of the solar electric process of converting sunlight directly into electricity. It also shows you that photovoltaics is a completely pollution-free way of generating power.

The working solar cell makes a fun gift for inquisitive children . . . of all ages.

To obtain your free cell, please fill out the coupon on the next page in BLOCK LETTERS and send it in an envelope to:

Solar Electric
175 Cascade Court
Rohnert Park, CA 94928
Tel.: (707) 586-1987

Yes, I would like to subscribe to *The Solar Electric Newsletter* and participate in the solar electric community around the world.

 I understand that my subscription entitles me to a 10 percent discount on all solar electric products sold by Solar Electric and that this offer comes with a full money-back guarantee. *Please rush me my first issue.*

☐ I enclose a check/money order for the sum of $25.00 made payable to Solar Electric. California residents please add 6 percent sales tax.

☐ I am paying by Visa/MasterCard/American Express.

My account number is: ⬚⬚⬚⬚⬚⬚⬚⬚⬚⬚⬚⬚⬚⬚

The card expires on: _____ / _____

SIGNATURE: _____

NAME: _____

STREET/BOX: _____

CITY: _____

STATE: _____ ZIP: _____

TELEPHONE: ( ) _____

—————————————— ✀ ——————————————

I would like to make use of your offer of a free working solar cell.

I purchased *The Solar Electric Book* on _____

(month/year) from: _____ at the

following address: _____

_____

I understand that I am entitled to only one free cell per copy.

I plan to use solar electricity for _____

I own: RV ☐ Boat ☐ Cabin ☐ Farm ☐ Airplane ☐ My age is: 18-22 ☐ 23-40 ☐ 41-61 ☐ 62 and over ☐

NAME: _____

STREET/BOX: _____

CITY: _____

STATE: _____ ZIP: _____

TELEPHONE: ( ) _____

☐ I am interested in receiving information on solar products.

☐ I am interested in receiving shareholder information about Solar Electric, which is a publicly owned California corporation.